The Minerals, Metals & Materials Series

The Minerals, Metals & Materials Series publications connect the global minerals, metals, and materials communities. They provide an opportunity to learn about the latest developments in the field and engage researchers, professionals, and students in discussions leading to further discovery. The series covers a full range of topics from metals to photonics and from material properties and structures to potential applications.

Ruslan Z. Valiev • Igor V. Alexandrov
Megumi Kawasaki • Terence G. Langdon

Ultrafine-Grained Materials

TMS

Springer

Ruslan Z. Valiev
Institute of Physics of Advanced Materials
Ufa University of Science and Technology
Ufa, Russia

Laboratory of Dynamics and Extreme
Characteristics of Promising
Nanostructured Materials
Saint Petersburg State University
St. Petersburg, Russia

Megumi Kawasaki
School of Mechanical, Industrial
and Manufacturing Engineering
Oregon State University
Corvallis, OR, USA

Igor V. Alexandrov
Department of Materials Science
and Physics of Metals
Ufa University of Science and Technology
Ufa, Russia

Terence G. Langdon
Materials Research Group
Department of Mechanical Engineering
University of Southampton
Southampton, UK

ISSN 2367-1181 ISSN 2367-1696 (electronic)
The Minerals, Metals & Materials Series
ISBN 978-3-031-31728-6 ISBN 978-3-031-31729-3 (eBook)
https://doi.org/10.1007/978-3-031-31729-3

This Springer imprint is published by the registered company Springer Nature Switzerland AG
The registered company address is: Gewerbestrasse 11, 6330 Cham, Switzerland

Preface

Bulk ultrafine-grained (UFG) materials produced via severe plastic deformation (SPD) have been the object of extensive research for the past two decades, and hundreds of scientific articles and several books have been published demonstrating the great prospects of these materials for fundamental studies and applied research. Although SPD processing is used for grain refinement in various crystalline materials such as composites, polymers, even ceramics, special attention is paid to metallic materials, metals and alloys. In recent years, there has been a breakthrough in this area, associated both with the development of new processing routes for the fabrication of bulk UFG metals and with investigations of the fundamental mechanisms that lead to novel properties for these materials. At the same time, there is by now no proper textbook on this subject for university/college students, postgraduates, and early-career scientists.

This book summarizes the main principles of SPD processing, describes the effects of SPD on grain refinement and phase transformations, and provides detailed overviews on the enhanced mechanical and functional properties of bulk UFG metals with respect to their innovative applications. These applications include nanostructured Ti-based materials in biomedical engineering, Cu and Al alloys in electrical engineering, nanostructured steels in construction engineering, and many others. Moreover, this book describes the application of severe plastic deformation for the formation of hybrid metal systems from simple powders and solid metals for an enhancement in the multifunctional properties.

The authors of this book have published pioneering works in the area of bulk UFG metals obtained via severe plastic deformation and they have made significant contributions in this field.

This textbook delivers the material in a logical way and accessible language with the main focus on key terminology and test questions and problems to ensure the understanding and learning of the material. We believe, therefore, that this book will become popular among experts as well as young researchers and students.

Ufa, Russia — Ruslan Z. Valiev
St. Petersburg, Russia — Igor V. Alexandrov
Ufa, Russia — Igor V. Alexandrov
Corvallis, OR, USA — Megumi Kawasaki
Southampton, UK — Terence G. Langdon

Acknowledgements

The use of bulk ultrafine-grained materials produced via severe plastic deformation (SPD) has remained an object of close interest and thorough research for more than two decades, prompting numerous scientific articles and several books but nevertheless lacking a solid guide to fundamental studies and applied research of these materials. Our understanding of these issues has been summarized in the present textbook which is designed to introduce the material in a logical way providing experts as well as young researchers and students with the opportunity to read, comprehend, and self-test their knowledge. We would like to express our sincere gratitude and appreciation to all who provided support, offered comments, discussed, allowed quoting of their remarks and publications, and otherwise assisted in the preparation of this textbook, particularly our reputable colleagues and friends from the International NanoSPD Steering Committee (www.nano-spd.org) and other members of the Materials Science community who have been actively working with us to publish many joint works.

The preparation of this textbook was made possible through the support of various funding agencies including the Ministry of Science and Higher Education of Russian Federation (Agreement No. 075-15-2022-1114 as of 06/30/2022 for RZV), the Russian Science Foundation under Agreement No. 22-19-00445 (RZV) and under Agreement No. 23-43-00041 (IVA), the National Science Foundation of the United States under Grant No. CMMI-2051205 (MK), and the European Research Council under ERC Grant Agreement No. 267464-SPDMETALS (TGL).

Contents

Chapter 1
Ultrafine-Grained Materials

Introduction

Although the mechanical and physical properties of any crystalline materials are determined by several factors, the average grain size of the material generally plays a very significant, and often a dominant, role. Thus, the strength of all polycrystalline materials is related to the grain size, *d*, through the famous Hall–Petch equation which states that the yield stress, σ_y, is given by

$$\sigma_y = \sigma_0 + k_y d^{-1/2} \tag{1.1}$$

where σ_0 is termed the friction stress and k_y is a constant of yielding. It follows from Eq. (1.1) that the strength increases with a reduction in the grain size and this has led to an ever-increasing interest in fabricating materials with extremely small grain sizes. In addition, the ideas and early experiments of Prof. Gleiter on nanocrystalline materials with a grain size of less than 100 nm and having a very high density of grain boundaries promoted approaches with emphasis on interface structure for the achievement of unusual and beneficial properties [1].

The grain sizes of commercial metals and alloys are generally tailored for specific applications by making use of predetermined thermomechanical treatments in which the alloys are subjected to specified regimes of temperature and mechanical testing. However, these procedures cannot be used to produce materials with submicrometer grain sizes because there is invariably a lower limit, of the order of a few micrometers, which represents essentially the minimum grain size readily attained using these procedures. Accordingly, attention has been directed toward the development of new and different techniques that may be used for the fabrication of ultrafine-grained materials with grain sizes in the submicrometer and the nanometer range.

To place this book in perspective, let us consider first a formal definition of UFG materials. With reference to the characteristics of polycrystalline materials, UFG materials are defined as polycrystals having very small grains with average grain

R. Z. Valiev et al., *Ultrafine-Grained Materials*, The Minerals, Metals & Materials Series, https://doi.org/10.1007/978-3-031-31729-3_1

sizes less than ~1 μm. For bulk UFG materials, there are the additional requirements of fairly homogeneous and reasonably equiaxed microstructures and with a majority of grain boundaries having high angles of misorientation. The presence of a high fraction of high-angle grain boundaries is important in order to achieve advanced and unique properties (see Chaps. 2 and 3). Ultrafine-grained materials can have a very fine grain size such as less than 100 nm, in which case they are typically designated nanocrystalline materials.

Two basic and complementary approaches have been developed during the last three decades for the synthesis of UFG materials and these are known as the "bottom-up" and the "top-down" approaches.

In the "bottom-up" approach, UFG materials are fabricated by assembling individual atoms or by consolidating nanoparticulate solids. The examples of these techniques include inert gas condensation, electrodeposition, ball milling with subsequent consolidation, and cryomilling with hot isostatic pressing, where cryomilling essentially denotes mechanical milling in a liquid nitrogen environment. More detailed information can be found in comprehensive reviews [1, 2]. In practice, these techniques are often limited to the production of fairly small samples that may be useful for applications in fields such as electronic devices but are generally not appropriate for large-scale structural applications. Furthermore, the finished products from these techniques invariably contain some degree of residual porosity and contaminations which may be introduced during the fabrication procedure. Recent research has shown that large bulk solids, in an essentially fully dense state, may be produced by combining cryomilling and hot isostatic pressing with subsequent extrusion, but the operation of this combined procedure is expensive and at present is not easily adapted for the production and utilization of structural alloys in large-scale industrial applications.

The "top-down" approach is different because it is dependent upon taking a bulk solid with a relatively coarse grain size and processing the solid to produce a UFG microstructure through heavy straining or shock loading. This approach avoids the small product sizes and the contamination which are inherent features of materials produced using the "bottom-up" approach and it has the additional advantage that it can be readily applied to a wide range of pre-selected alloys.

The first observations of the production of UFG microstructures using the "top-down" approach appeared in the scientific literature toward the end of the 1980s and early 1990s in several publications dealing with pure metals and alloys [3]. It is important to note these early publications provided a direct demonstration of the ability to employ heavy plastic straining in the production of bulk materials having fairly homogeneous and equiaxed microstructures with grain sizes in the submicrometer range and with a high fraction of high-angle grain boundaries.

In order to convert a coarse-grained solid into a material with ultrafine grains, it is necessary both to impose an exceptionally high strain in order to introduce a high density of dislocations and for these dislocations to subsequently rearrange to form an array of grain boundaries. In practice, however, conventional metal-working procedures, such as extrusion or rolling, are restricted in their ability to produce UFG structures for two important reasons. First, there is a limitation on the overall strains that may be imposed using these procedures because the processing techniques incorporate corresponding reductions in the cross-sectional dimensions of the work-pieces. Second, the strains imposed in conventional processing are insufficient to introduce UFG structures because of the generally low workability of metallic alloys at ambient and relatively low temperatures. As a consequence of these limitations, attention has been devoted instead to the development of alternative processing techniques based on the application of severe plastic deformation where extremely high strains are imposed at relatively low temperatures without incurring any concomitant changes in the cross-sectional dimensions of the samples.

Formally, processing by severe plastic deformation (SPD) has been defined as those metal forming procedures in which a very high strain (typically with true strain $e \geq 6$–8) is imposed on a bulk solid without the introduction of any significant change in the overall dimensions and leading to the production of exceptional grain refinement so that, usually, the processed bulk solids have 1000 or more grains in any section [3, 4].

In principle at least, the use of heavy deformation in metal processing, with the objective of producing metal alloys with improved properties, has a long history which may be traced back to the early metal-working of ancient China, the blacksmith's production of high-quality Damascus steel in the Middle East and the fabrication of the legendary Wootz steel in ancient India (see [4], for example). However, from a modern scientific perspective, developments in processing by SPD were first stimulated over 60 years ago through the extensive and meticulous work of Prof. Percy Bridgman at Harvard University. Bridgman conducted a remarkably comprehensive series of experiments on the application of high pressures to bulk solids and, in 1952, the results from these many experiments were succinctly summarized in a book [5]. It is important to note that Bridgman received the Nobel Prize in Physics in 1946 with the citation reading "for the invention of an apparatus to produce extremely high pressures, and for the discoveries he made therewith in the field of high pressure physics." Bridgman was the first to propose the device for processing of metals through a combination of compression and torsional straining. This device was further developed in several laboratories and ultimately used as the SPD procedure later named as high-pressure torsion (HPT) (see Chap. 1).

A second major influence on SPD processing may be traced to the pioneering work of Dr. V.M. Segal and his colleagues, conducted in the 1980s in Minsk in the former Soviet Union (now the capital of Belarus). Segal and coworkers were the first to develop the technique named equal-channel angular pressing (ECAP) (or equal-channel angular extrusion), and this process was developed by Valiev et al. in the 1990s as an SPD procedure for processing of UFG materials where this technique is now used very widely.

Since this early work, several different SPD techniques have been proposed, developed, and evaluated. These techniques include modifications of ECAP and HPT as well as other techniques—for example, multidirectional forging, twist extrusion, cyclic extrusion–compression, reciprocating extrusion, repetitive corrugation and straightening (RCS), constrained groove pressing (CGP), cylinder covered compression (CCC), accumulative roll-bonding (ARB), friction stir processing

(FSP), and others (see [2, 6]. All of these procedures are capable of introducing large plastic straining and significant grain refinement in bulk crystalline solids, while reasonably retaining the original dimensions of the workpieces. Some of these techniques, such as ECAP, HPT, multidirectional forging, and ARB, are already well-established methods for producing UFG materials where, depending upon the crystal structure, the processed microstructures have grain sizes lying typically within the range of ~70–500 nm. Some other techniques are also currently under development for this purpose.

Already, in early studies, it was found that the SPD processing of metallic materials leads to strong grain refinement as well as to unusual phase transformations [6] as the dissolution of secondary phases, the accelerated decomposition of solid solutions, often to amorphization, and even dynamic aging. Subsequently, these problems were the subject of numerous studies and it was demonstrated in particular that the kinetics of these transformations is closely related to the SPD processing regimes—specifically, the temperature, amount, and rate of strain and the applied pressure. At the same time, deformed UFG structures and the development of phase transformations due to severe deformation led to a high density of various nanostructured features in materials, the formation of different types of grain boundaries, including nanotwins, the appearance of nanosized precipitations of secondary phases, the formation of grain boundary precipitates, etc. Such nanostructuring allows for the production of materials with novel properties, including superior multifunctional properties where there is a combination of high mechanical (very high strength and ductility, superplasticity) and functional properties (corrosion and radiation resistance, electrical conductivity, etc.) that are induced by nanostructural design. Several examples of the use of multifunctional nanomaterials for their innovative applications in medicine and engineering are considered and discussed in detail in this book.

The scientific and methodological sections of the textbooks are provided by questions and exercises.

1.1 Principles of SPD and Processing Techniques

The formation of ultrafine grains, characterized mainly by high-angle grain boundaries, becomes possible as a result of processing using SPD. However, it is necessary to fulfill a number of conditions related to the choice of SPD regimes [6].

The first of these conditions is to provide low deformation temperatures. It is known that a reduction in the temperature of plastic deformation slows down the course of diffusion processes that are responsible for recovery and recrystallization in the microstructure. As a result, the defect density in the deformed material increases, and the reduced temperatures of SPD ($T < 0.4\ T_m$) provide a potential for achieving extremely high dislocation densities up to 10^{14} m^{-2} and even higher. The ultimate values are considered to be 10^{16}–10^{17} m^{-2}. The self-organization of dislocation ensembles leads to the formation of low-angle and then high-angle grain boundaries having a high density of extrinsic grain-boundary dislocations. The latter are responsible for the appearance of long-range fields of elastic stresses extending far into the grain bodies and affecting the level of various properties of bulk nanostructured materials. In practice, the intragranular regions (grain bodies) are practically free of dislocations.

The second condition is to achieve high accumulated strains (true strain $e > 6$–8). With such strains, a typical cellular structure with low-angle misorientations fully develops into a grain structure with high-angle grain boundaries and a balance is ultimately established between the rates of accumulation and annihilation of dislocations.

The third condition is to implement the deformation process at high applied pressures (usually above 1 GPa). An increase in the applied pressure leads to an increase in the deformation ability of the material and ensures preservation of the integrity of a deformed workpiece. High pressures hinder the diffusion processes and inhibit the processes of annihilation of deformation defects, which self-organize into low-angle and gradually into high-angle grain boundaries. Furthermore, these processes occur at lower strain degrees.

The fourth condition is to ensure a turbulent non-monotonic nature of material flow in the volume of a deformed workpiece at the macro level and an active displacement and rotation of grains during SPD. In this case, the processes of grain boundary sliding and grain boundary diffusion can take place. New slip and/or twinning systems are involved in the deformation process.

Finally, the fifth condition contributing to the achievement of an increased dislocation density and a smaller grain size is to use alloys with an ordered structure and materials with a low stacking fault energy.

Special deformation methods need to be developed and applied to fulfill the above conditions aimed at an implementation of severe plastic deformations and the formation of ultrafine-grained states with predominantly high-angle grain boundaries.

Conventional deformation modes, such as tension, compression, drawing, rolling, and others, lead to a significant shape change in deformed workpieces, and typically the accumulated strains fail to exceed ~2–3. As a result, a subgrain structure with predominantly low-angle misorientations is formed.

Special deforming techniques need to be applied to implement the above-mentioned conditions of SPD. The geometric sizes and shape of a workpiece, defined as the scale factor, also significantly affect the uniformity of deformation. The friction between the deformed workpiece and the die set should be taken into account and a special lubricant used to reduce the friction.

Let us consider some of the most popular SPD techniques that satisfy the above requirements [6]. They are high-pressure torsion, equal-channel angular pressing, multidirectional forging and some others.

One of the first techniques, with the help of which the SPD principle was implemented, was high-pressure torsion (HPT). The implementation of this technique is based on a combination of shear deformation and high applied quasi-hydrostatic pressures.

Fig. 1.1 Professor Bridgman conducts experiments on high pressure studies [7]. Image courtesy of Smithsonian Institution Archives

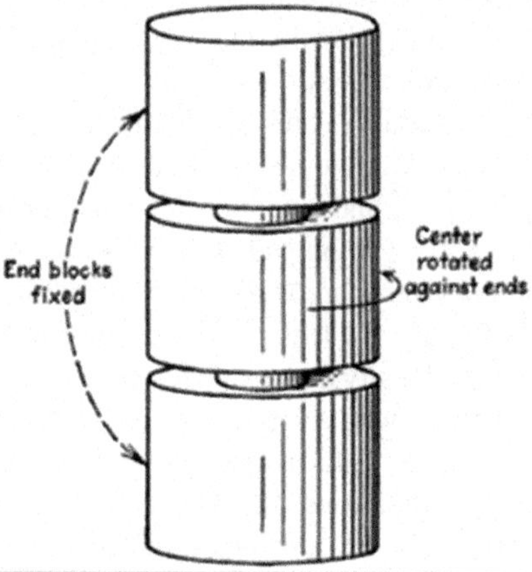

Fig. 1.2 Bridgman experiment for compression and torsion test [5]. (Reprinted from [8], Copyright 2008, with permission from Elsevier)

As mentioned above, the pioneering studies of the effect of high pressures on material behavior were carried out by Percy Williams Bridgman (Fig. 1.1) [7]. In 1946, he became the Nobel Prize Winner "for the invention of a device that allows creating ultrahigh pressures, and for the discoveries made in connection with this in high-pressure physics." Bridgman used pressures from 400 MPa to 4000 MPa, which are about 4000–40,000 times higher than the atmospheric pressure.

Figure 1.2 displays a scheme of the Bridgman experiment, during which the central part of the workpiece was rotated under the action of a torque relative to its upper and lower parts in the absence or presence of axial compressive stresses [8].

The regularities established by P.W. Bridgman were developed as the SPD technique named "high-pressure torsion." This term was introduced at the first International Conference on the NanoSPD subject area organized in Moscow in 1999 [9]. The technique envisages the use of disc-shaped samples that are placed between two anvils under high applied pressures of several GPa (Fig. 1.3) [10]. The applied compressive pressures result in static friction forces. As a result, it becomes

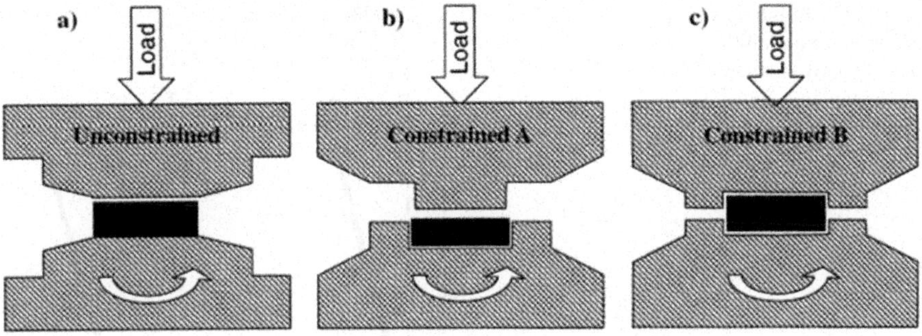

Fig. 1.3 Principles of HPT for (**a**) unconstrained and (**b**, **c**) two types of constrained conditions. (Reprinted by permission from Springer Nature [10], Copyright 2007)

possible to avoid slipping on the contact surfaces between the disc and the anvils during rotation of one of the latter. The same frictional forces cause the workpiece to be deformed by shear up to a very high strain.

To perform shear deformation under the conditions of quasi-hydrostatic compression, the workpiece can be partially or completely placed in the cavities made in the anvils and fitting the workpiece size [6]. The material does not essentially flow out of the space between the anvils.

The true accumulated strain of the workpiece during high-pressure torsion is calculated using the expression [6]

$$\varepsilon = \ln\left(\frac{\varphi \cdot r}{h}\right) = \ln\left(\frac{2\pi N \cdot r}{h}\right) = \mathrm{tg}\,\gamma(r) \tag{1.2}$$

where φ is the angle of relative rotation of the anvils, N is the number of rotations, r is the distance from the rotation axis passing through the workpiece center, perpendicular to its surface, h is the workpiece thickness, and γ is the degree of shear strain (Fig. 1.4) [11].

As follows from expression (1.2), the true strain is higher when there are higher numbers of full rotations of one of the anvils, longer distances from the axis of rotation to the point at which the strain degree is calculated, and smaller thicknesses of the workpiece. In this case, the diameter of the workpieces subjected to HPT is, as a rule, from 10 to 20 mm, and the thickness does not exceed 1 mm.

It follows from expression (1.2) that the true strain depends on the distance to the axis of rotation. It has been demonstrated that strain turbulence in the case under consideration leads to the leveling of differences in the microstructure and properties as the strain degree increases. Since the workpiece does not change its geometric dimensions during high-pressure torsion, the deformation can continue until very large strain degrees are achieved after several rotations.

As mentioned above, in 1988, high-pressure torsion was for the first time demonstrated by Valiev et al. as a technique for the formation of an ultrafine-grained

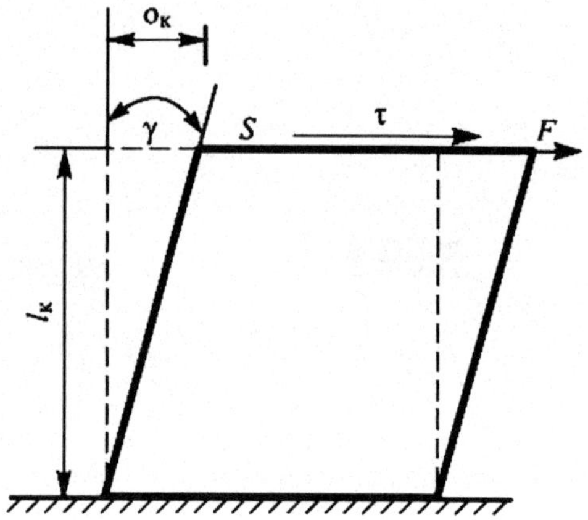

Fig. 1.4 Parameters of shear deformation [11]. From GG Bondarenko et al., Fundamentals of materials science 3 ed, 2020. With permission from BKL Publishers

structure with mainly high angle boundaries, providing the appearance of new properties [3]. This work effectively initiated the research age of NanoSPD.

In general, the high-pressure torsion technique is presently used primarily for scientific purposes, allowing various fixed strain degrees to be achieved. Deformed samples tend to be very restricted in size. At the same time, when implementing this method, very high pressures of several GPa can be applied to workpieces under deformation.

An increase in the strain degree and applied pressure favorably affects the formed nanostructured states, leading to a decrease in grain size, a growth of grain boundary misorientations, the formation of a more homogeneous microstructure, and an increase in microhardness.

Over the last several years, the high-pressure torsion technique has been developed to achieve high strains during the deformation of relatively massive samples with shapes close to cylindrical and for wires as a continuous process.

Equal-channel angular pressing (ECAP) is another common technique for producing bulk nanostructured billets using SPD.

This method was proposed by V.M. Segal and his staff as a way to implement the deformation of massive workpieces with simple shear. In the beginning of the 1990s, R.Z. Valiev and colleagues demonstrated that this technique allowed the production of bulk billets with ultrafine grains and attractive properties from various metallic materials [3].

During ECAP, a massive billet is pressed multiple times through two intersecting channels (Fig. 1.5), where the channels have the same cross-sections. To reduce the friction between the channel walls and the deformed workpiece, some lubricant is typically used.

The square die set for ECAP can be disassembled and consists of two parts with a bolted connection. In this case, a channel with a square cross-section is made in one of the parts. In the nondisassembled cylindrical die set, intersecting cylindrical channels are drilled. The design of a square die set is simple, but there is a

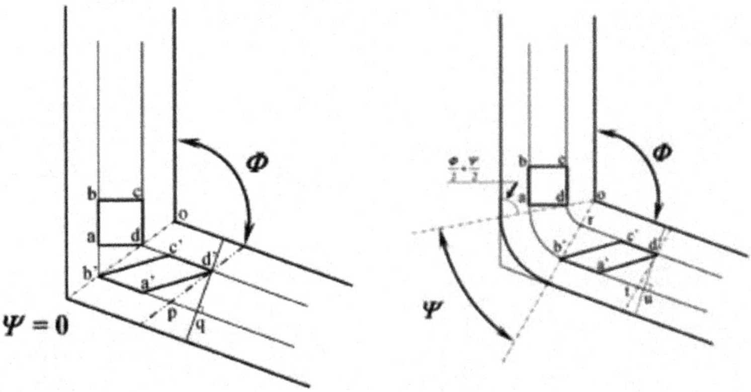

Fig. 1.5 Scheme of a die-set for ECAP. (Reprinted from [12], Copyright 1996, with permission from Elsevier)

possibility of extrusion of material of the deformed workpiece through the plane of the two parts of the die set outside of the channels.

Figure 1.5 displays a die set scheme for ECAP. The angle Φ shown in the scheme is the angle of channel intersection, thus designated the internal angle. The angle Ψ is the external angle of channel intersection [6, 12].

The workpiece is placed in a vertical channel, pressed by a punch and emerges from the near-horizontal channel. As can be seen in Fig. 1.5, the square element *abcd* in the flow plane changes its geometry in the area of the channel intersection, which is conditioned by the shear strain experienced by it. Moreover, the degree of shear strain γ depends on the value of both angles in accordance with the expression [6]

$$\gamma = 2\cot\left(\frac{\Phi}{2}+\frac{\Psi}{2}\right)+\psi\cos ec\left(\frac{\Phi}{2}+\frac{\Psi}{2}\right). \tag{1.3}$$

The greatest attention of researchers is drawn to die sets with an internal angle of channel intersection equal to 90°. If the external angle is $\Psi = 20°$, then each pass of the workpiece through the die set will lead to an additional equivalent strain

$$\varepsilon = \frac{\gamma}{\sqrt{3}} \tag{1.4}$$

which is equal approximately to 1.

The area of the equipment in which shear deformation occurs is called the deformation zone. A decrease in the external angle makes the deformation zone narrower and concentrates the shear deformation near the plane of the channel intersection.

Near the external arc, tracing out the profile of intersecting channels and the external angle Ψ, a so-called "dead zone" can be formed which is not filled with the material of a deformed workpiece. The volume of the "dead zone" increases with the hardening of the material of a deformed workpiece. It can be reduced either by

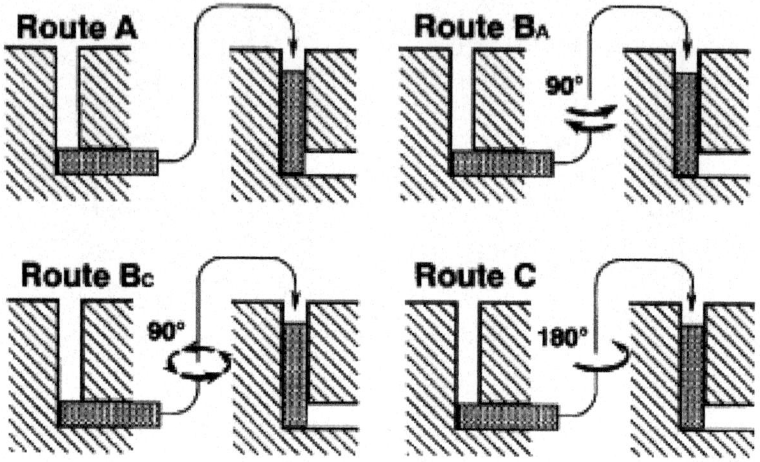

Fig. 1.6 ECAP routes. (Reprinted from [13], Copyright 2000, with permission from Elsevier)

varying the die set geometry by changing the external angle Ψ, or through the use of back pressure in the horizontal channel of the ECAP die set.

A workpiece can be pressed through the die set repeatedly. Successful extraction of the deformed workpiece out of the die set is carried out by subsequent pressing of the next workpiece through the die.

Between successive passes through the ECAP die set, the orientation of a workpiece may be changed (Fig. 1.6) [13]. The following ECAP routes are identified: route A—the workpiece orientation does not change between successive passes; route B_A and B_C—the workpiece is rotated at 90° about the longitudinal axis in the same direction and in opposite directions, respectively, between successive passes; route C—the workpiece is rotated through an angle of 180° relative to the longitudinal axis between successive passes.

The use of various ECAP routes permits controlling the deformation processes occurring in a workpiece and ensuring the constancy of the direction of shear deformation or its targeted change. The degree of accumulated strain increases in multiples of the number of passes applied to the workpiece through the die set. As a rule, after four ECAP passes, an ultrafine-grained state is formed, but this may be improved during subsequent ECAP passes. A further increase in the number of passes (up to 12–16) leads to an improvement in the UFG structure with saturation. The grain size is then stabilized, the average misorientation angle between neighboring grains increases slightly and the microstructure becomes more homogeneous.

When implementing the ECAP process, self-heating of a workpiece should be considered. An increase in a workpiece temperature is associated with energy dissipation during deformation and can reach several tens of degrees. The degree of self-heating of alloys is higher than that of pure metals.

By contrast, the speed of ECAP implementation has little effect on an emerging UFG structure.

ECAP is successfully used to form UFG structures in massive billets of various pure metals and alloys. The length of workpieces is usually from 60 to 300 mm, and the diameter of the cross-section or the rib is from 10 to 50 mm.

In some cases, such as for refractory metals and alloys, additional heating of a die set and a workpiece is required for successful implementation of ECAP. A die set of tool steel withstands heating up to temperatures not higher than 450–500 °C. In this case, the workpiece can be heated to higher temperatures. To prevent oxidation processes, the workpiece may be placed in an envelope of ductile material.

The *multidirectional forging* technique is based on multiple repetitions of free-forging operations: upsetting–broaching with a change in the axis of the applied deforming force. This method can be applied using standard industrial press equipment; however, the uniformity of deformation in this technological scheme is lower by comparison with the schemes of ECAP and HPT. The multidirectional forging treatment begins at elevated temperatures of about 0.6 T_m and ends near 0.3 T_m. At the same time, low specific loads on a tool are ensured, and it becomes possible to form UFG states in hard-to-deform materials.

On the whole, it should be noted that the development of techniques for practical implementation of the SPD processing is in progress. Special attention is paid to the transition from laboratory to industrial conditions in the production of bulk UFG billets [6].

1.2 Grain Refinement, Its Mechanisms, and Ultrafine Grains

An increase in the plastic strain degree is accompanied by microstructure evolution. One of the main processes of this evolution is the formation of new interfaces, leading to grain structure refinement.

In recent decades, numerous experimental studies on microstructure evolution during large plastic deformations of metallic materials have been carried out using traditional processing schemes such as rolling, extrusion, etc., where true strain degrees e typically reach values of ~3 ÷ 4. Various pure metals with fcc, hcp, and bcc lattices were deformed. Room temperature and, in some cases, lower temperatures were chosen as the deformation temperatures. The most detailed studies were carried out by V.V. Rybin et al. and N. Hansen et al. in the 1980–1990s (see, e.g., [14–18]).

When analyzing the processes of microstructure evolution during large deformations, attention was paid to such microstructure parameters as grain and cell sizes, the misorientation angle between them, the interface thickness and the dislocation density. The information about these parameters was obtained using optical microscopy, transmission electron microscopy, scanning electron microscopy, and X-ray diffraction analysis.

According to Rybin, there are two stages of large deformations. At the first stage ($0 < e < 0.3$), an increase in the density of dislocations is observed and dislocation cells are formed. At the second stage ($e > 0.3$), called the stage of developed

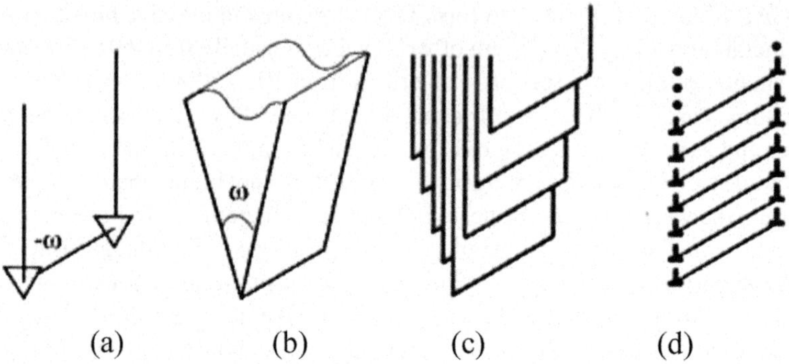

Fig. 1.7 Relation between wedge disclinations and terminated edge dislocation walls: negative wedge disclination with the strength x (**a**), wedge of angle x which has to be inserted to create a disclination (**b**), a set of half-planes modeling the wedge (**c**), terminated wall of equidistant dislocations (**d**). (Reproduced from [15], Copyright 2004, with permission of John Wiley & Sons)

deformation, the size of dislocation cells practically does not change, but adjacent regions with sizes exceeding those of cells turn against each other. Translational shears and plastic turns occur as a result of the activation of collective modes of dislocation motion, described as nucleation and motion of partial disclinations that are carriers of rotational plasticity modes (Fig. 1.7). Partial disclinations are abrupt walls of edge dislocations, which ensure a larger turn between neighboring regions in the crystal, as there is a larger number of dislocations in the wall, equal to the power ω of partial disclination.

The performed experimental studies show that a cellular dislocation structure is formed inside grains of the initial coarse crystalline microstructure in metals with an fcc lattice at the initial stages of plastic deformation. The misorientation angle between adjacent cells is low, and it is 0.1–0.2°. As the deformation process develops, dislocation cells remain weakly misoriented. In addition, the formation of dislocation walls is observed inside the cellular structure forming abrupt boundaries misorienting neighboring regions by angles exceeding 1° (Fig. 1.8).

As the deformation develops, the microband boundaries are refined and sharpened. The numbers of such boundaries and the corresponding misorientation angles are growing, with the latter reaching values of 8–10°. The orientation of microbands in relation to the rolling direction (RD) also changes. The microbands rotate from a position characterized by an angle of about 35° with respect to RD, toward RD (Fig. 1.9). In this case, the microbands practically do not change their sizes. At the same time, the microbands are combined into mesobands extending over great distances along the rolling plane. After the entire volume of the deformed material is filled with microbands, further deformation is carried out by nucleation of shear bands oriented at an angle of 35° to the mesobands and inside the mesobands. Shear bands are highly fragmented. There are cells inside them the size of which changes slightly with the development of deformation.

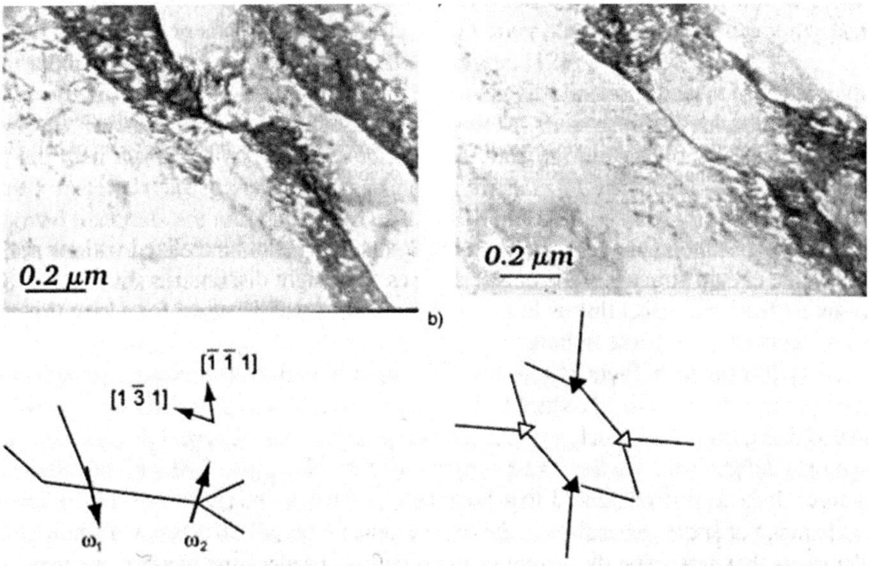

Fig. 1.8 Formation of abrupt boundaries in Cu single crystal rolled down to 70% thickness reduction at room temperature. ω_1 and ω_2 – powers of junction disclinations. Filled and empty triangles – positive and negative disclinations, respectively. (Reproduced from [15], Copyright 2004, with permission of John Wiley & Sons)

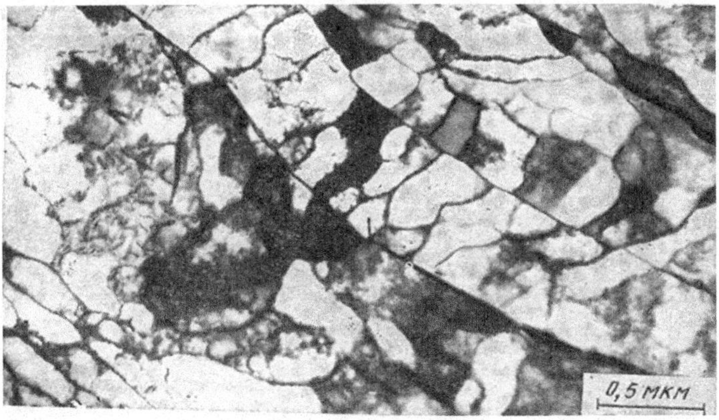

Fig. 1.9 Dipole configuration of microband boundaries in Ni ($e = 1.05$) [16]

The processes of microstructure evolution at large deformations in metals with hcp and bcc lattices generally correspond to those that occur in metals with an fcc lattice, but they have some important features.

Deformation twins can play an important role in the implementation of large deformations in metals with an hcp lattice. The microstructure of metals with a bcc lattice is characterized by closed loop-like and dipole configurations of abrupt dislocation boundaries, band structures, and the branching of abrupt dislocation boundaries. The characteristic dimensions of these structural components are 0.2–0.5 μm and the misorientation angles between neighboring regions are 1–5°.

When analyzing the processes accompanying large deformations, N. Hansen and colleagues focused on the transformation of a cellular microstructure into a granular one.

According to their observations, a characteristic feature of the microstructure at relatively low degrees of compression (about 70%) is the presence of boundaries directed at an angle to RD (Fig. 1.10), as well as lamellar boundaries, almost parallel to the direction of rolling. For large degrees of compression (90–98%), lamellar boundaries are typical (Fig. 1.10).

The localized shear regions in the form of shear bands (*SB* in Fig. 1.10) are also typical of such structures. These shear bands are intense shear regions that span several grains.

The main element of the mesoscale structure is cell blocks. At low strain degrees, the cell blocks are bounded by dense dislocation walls (DDWs) which are extended almost flat boundaries of a deformation origin. At the boundaries of cell blocks,

Fig. 1.10 Ni. Flat rolling. Compression degree ε = 98%. TEM. Longitudinal section containing rolling direction. (Reprinted from [17], Copyright 2000, with permission from Elsevier)

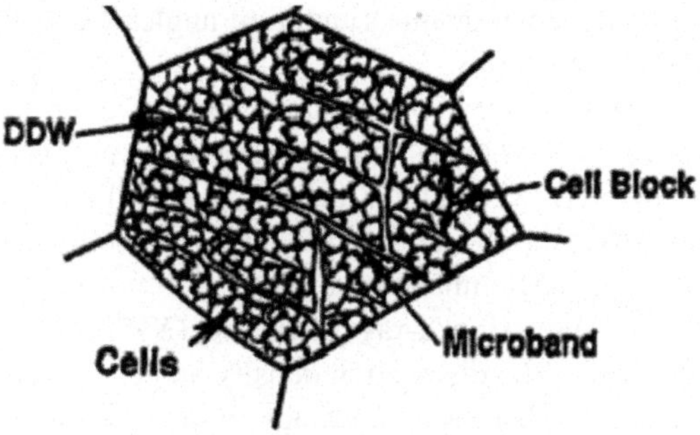

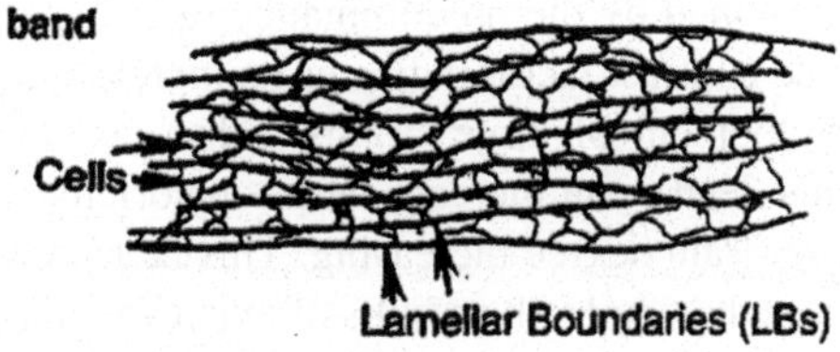

Fig. 1.11 Scheme of deformation microstructures and division of grain according to the data of N. Hansen: (**a**) low and moderate strains; (**b**) large plastic strains. (Reprinted from [17], Copyright 1997, with permission from Elsevier)

microbands can be found and these are flat areas between two closely located DDWs (Fig. 1.11a).

At high strain degrees, elongated cell blocks in the formed bamboo structure are separated by almost flat extended lamellar boundaries (LBs), apparently separating adjacent bands of intense shear.

Cell blocks are made up of individual cells. Inside the cells the dislocation density is very low. With large deformations, the cell boundaries intersect the short space between the lamellar boundaries in the bamboo structure (Fig. 1.11b). The cell boundaries are incidental dislocation boundaries (IDBs) formed as a result of the capture of gliding dislocations.

With an increase in the strain degree, the sizes of the blocks of cells and the cells themselves are reduced. The size of the first decreases faster than the size of the second, and as a result, the number of cells in the block decreases. At high strain degrees, the boundaries of cell blocks are oriented almost parallel to the rolling plane.

When analyzing the microstructure of highly deformed metallic materials, the geometrically necessary boundaries (GNBs) and IDBs are usually identified. GNBs include extended boundaries of cell blocks (DDWs, microbands, and LBs). These boundaries divide the regions in which various combinations of slip systems operate and the strain value is different. IDBs are common cell boundaries that form as a result of the interaction of gliding dislocations. From the morphological point of view, GNBs include those boundaries along which at least three cells are located. In addition, the boundaries of those subgrains are characterized by equilibrium triple junctions and/or misorientation angles exceeding 15° which are considered to be such boundaries.

An increase in the compression degree from 70% to 98% during rolling of Ni is accompanied by a slight increase in the number of GNBs with high-angle misorientations. However, their number, even with a maximum compression degree of 98%, is still insignificant, and the overwhelming majority of GNBs in the microstructure has low-angle misorientations.

The precision X-ray diffraction (XRD) data indicate a significant decrease in the growth of the dislocation density and even its constancy at the stage of large deformations. There is a constancy in the size of coherent scattering domains (cells) or, at least, the reduction rate of their size decreases significantly. As the deformation develops, the weighted dislocation contrast factor changes, the value of which is determined by the predominance of either edge or screw dislocations. The change in the contrast factor indicates an increase in the share of edge dislocations with the development of large strains. Edge dislocations, self-organizing into dislocation walls, lead to an increase in the misorientation angle of neighboring cells with the large strain degree increasing. This fact is confirmed experimentally.

On the whole, the results of experimental studies testify to significant changes in the microstructure of metallic materials during large deformations (up to $e = 1–2$). Fragmented structures form in the deformed metal. An increase in the strain degree leads to a decrease in the fragment size, but the misorientation angles grow slightly.

In the late 1980s and early 1990s, when the first fabrication of UFG materials, using high pressure torsion and then ECAP was demonstrated [3] (Fig. 1.12), research in this promising scientific field has been actively developed and has attracted wide international interest due to numerous fundamental aspects and the potential advanced properties inherent in UFG metal materials processed by SPD

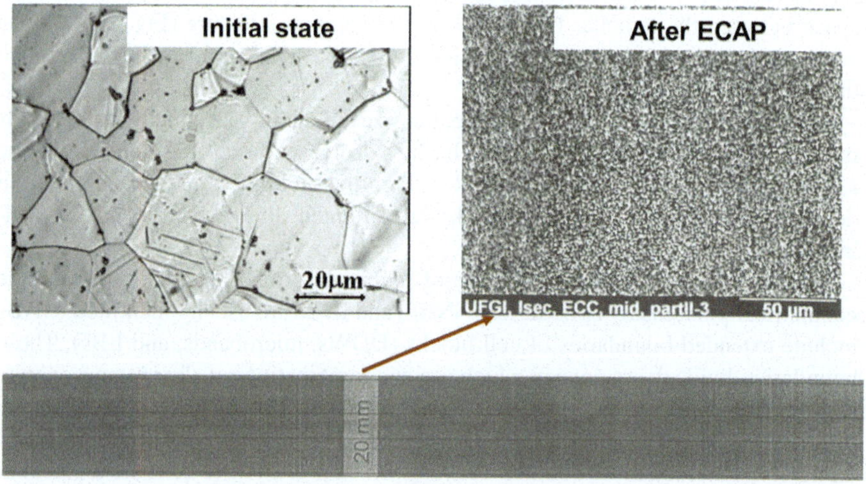

Fig. 1.12 The early observation of UFG structure formation in copper billet by ECAP, eight passes. (Reprinted from [3], Copyright 2000, with permission from Elsevier)

methods. The features of microstructures and the promising properties of these materials are described in the following chapters of this textbook.

1.3 Nanostructural Architecture of SPD-Processed Materials

Studies performed during the last decade, conducted using a set of advanced techniques, including 3D-atom probe, TEM/HREM, X-ray diffraction, and others, reveal that SPD processing can not only form UFGs but also it may be used to engineer grain boundary structures and generate nanoparticles, segregations, or nanoclusters and other structural elements at the nano-scale. The nature of these phenomena is related to the fact that, together with grain refinement, processing by SPD may also lead to unusual phase transformations as a dissolution of second phases, precipitation, amorphization, allotropic phase transformations, and other processes, producing various nanostructural features (see, for example, [19–22]). Also, the unusual combinations of SPD-induced diffusive and displacive (martensitic) phase transformations (i.e., transitions with or without mass transfer) can take place during SPD [23–25]. On the other hand, the type and morphology of such nanostructural elements, as well as their number density, control deformation and transport mechanisms responsible for improvements of the mechanical, chemical, and physical properties of bulk nanostructured materials through SPD techniques and the optimization of processing regimes. These issues on nanostructuring from SPD processing are considered in this section.

1.3.1 Grain Boundaries in Nanostructured Materials

The density of grain boundaries (GBs) and interphase boundaries (IBs) in UFG materials with submicron (100–1000 nm) or nanocrystalline (<100 nm) sizes is very high, and they can make a significant impact on the novel level of properties. This justifies the reason for considering UFG materials as interface-controlled materials [1, 26].

Gleiter and colleagues already in their first publications on nanocrystalline materials suggested that the atomic structure of grain boundaries in UFG materials may be characterized by a number of special features in contrast to grain boundaries in conventional polycrystalline materials [26]. It is especially important for materials in which UFG microstructures are produced by SPD techniques [3, 19, 22].

Although the physical mechanisms of microstructure evolution during SPD processing call for further investigation, numerous studies convincingly demonstrate that the formation of a UFG structure is determined by the evolution of cells or dislocation substructures formed in the early deformation stages into ultrafine grains with high-angle boundaries during subsequent deformation. At the same time, and depending on the regimes of SPD processing, different types of grain boundaries can be formed in the UFG materials, such as high and low-angle, special and random, equilibrium and so-called "non-equilibrium" grain boundaries with strain-distorted structures containing extrinsic dislocations, as well as boundaries with GB segregations of alloying elements [6, 22].

As discussed in Sect. 1.2, the fabrication of UFG metals and alloys with predominantly high-angle grain boundaries by SPD techniques was revealed already in early works by Valiev et al. [3, 4, 27]. In the recent decade, active development and application of the electron back scattering diffraction (EBSD) technique for structural analysis of materials has provided quantitative data on the low- to high-angle GBs ratio and the presence of special and random boundaries in various metals and alloys subjected to SPD.

For example, back-scattered electron image analysis has been used to plot carefully misorientation histograms [28]. A Mackenzie plot (i.e. the probability density of observing a particular disorientation angle between randomly oriented cubic crystals) has been added to each histogram for an ECAP-processed Al 6016 alloy. The boundary misorientation angle histogram in Fig. 1.13 shows that after the first ECAP pass the microstructure is characterized by ~93% of the boundaries with misorientation angles of less than 15° confirming that a large population of low-angle boundaries is characteristic for this state.

After the second pass (Fig. 1.13b), the average dislocation cell size is reduced considerably, but the inhomogeneous structure remains characteristic for most cells with the Y-(or transverse) direction. Simultaneously the boundary misorientation distribution shifts upward, with ~67% of boundaries comprising misorientation angles of less than 15°. After four ECAP passes, some approximately equiaxed

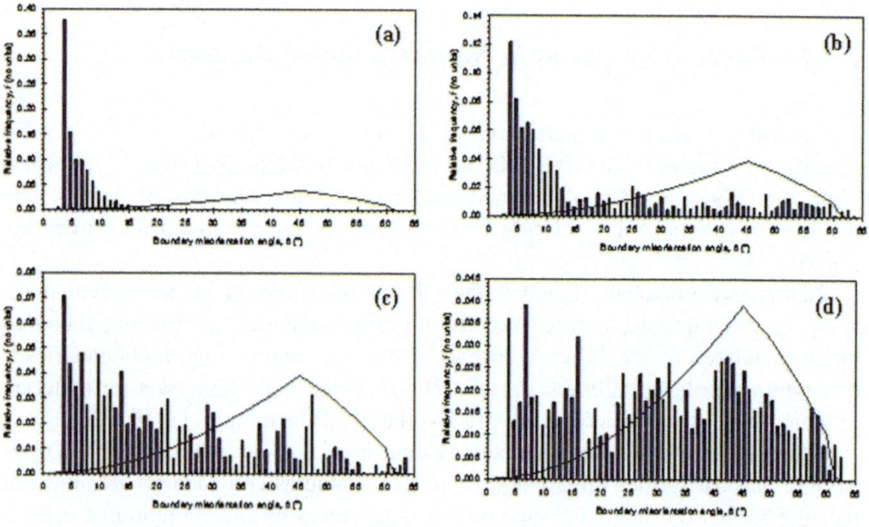

Fig. 1.13 Relative frequency of dislocation cell/subgrain boundaries as a function of misorientation angle for Al 6016 samples processed by ECAP. (**a**) One pass, (**b**) two passes, (**c**) four passes, and (**d**) eight passes. (Reprinted from [28], Copyright 2010, with permission from Elsevier)

dislocation cells/subgrains appear together with some shape irregularities and alignment of the microstructure. The distribution with ~34% of boundaries having a misorientation of less than 15° histogram reflects a more gradual upward shift in the corresponding misorientation angle (Fig. 1.13c). Further processing to 8 and 12 ECAP passes resulted in an evident reduction of the dislocation cell/subgrain size, leading to shifts in the misorientation distributions (Fig. 1.13d) and a near random distribution after 12 ECAP passes.

The results of this work, as well as another recent study by EBSD analysis [29], provide convincing support for earlier conclusions that the formation of a UFG structure with mostly high-angle boundaries takes place during ECAP when the number of passes is over 4–6 (true strain $e > 4–6$). Such very large strains cannot be achieved with the use of conventional processing techniques, such as rolling, extrusion or drawing, and therefore their application provides no opportunity for the formation of a real UFG microstructure with high-angle grain boundaries. Moreover, SPD techniques are related to the schemes of simple shear, at the forefront of which are the rotary modes, which makes it possible to form the ultrafine grains as equiaxial or at least close to it in shape.

A variety of structural methods was used in different studies to investigate features of grain boundaries in UFG materials produced by SPD techniques. Among them were methods of high resolution transmission electron microscopy (HRTEM), X-ray diffraction, Mössbauer spectroscopy, dilatometry, differential calorimetry, and others (see, for example, [3, 6, 30]). It was clearly demonstrated that optimization of the SPD processing regimes results in the formation of mostly high-angle grain boundaries with specific non-equilibrium structures. The application of structure-sensitive probes that had been performed to modifications of the atomic structure, such as grain boundary diffusion measurements [31–33] and HRTEM analyses [31, 34, 35], helped to identify and characterize the effect of the severe deformation processing on transformations of the grain boundary structures.

For example, it was revealed that HPT may result in the formation of an excessively high density of dislocations, facets, and steps, which were observed at grain boundaries of the UFG Al-3% Mg alloy, already in an earlier publication [36]. There are numerous reports which confirm the typical appearance of such non-equilibrium grain boundaries in different materials processed by SPD [3, 22, 33, 37].

The term "non-equilibrium" grain boundaries has been accepted and utilized by the entire SPD materials science community and it is also used here to describe grain boundaries which are defined formally as boundaries with strain-distorted structures and containing high densities of extrinsic dislocations [3, 22].

Thus, the recently conducted studies of the grain boundary structure in SPD-processed UFG materials give a possibility to draw the conclusion that specific non-equilibrium grain boundaries with strain-distorted structures associated with a correspondingly large residual microstrain is a typical feature in these nanomaterials.

1.3.2 Nanotwins in Metallic Nanomaterials

As was well established, decreasing the grain size usually impedes twinning in coarse-grained fcc metals. In the case of related nanocrystalline metals, the tendency is more complicated. In fact, twinning then becomes easier with decreasing grain size up to a certain grain size, reaching a maximum twinning probability, and then twinning becomes more difficult when the grain size decreases even further [38]. The difference between twinning mechanisms in nanocrystalline and coarse-grained materials has been revealed by means of molecular dynamics simulations and experimental observations. Consequently, deformation twins of several types are observed in nanocrystalline materials only, but not in coarse-grained metals. The formation of nanotwins can be promoted if some properties intrinsic to the material and conditions of external deformation are satisfied. Among these properties are: (1) a relatively low stacking fault energy; (2) a low deformation temperature; and (3) a high strain rate [38]. It seems that there is a range of grain sizes which is optimal for deformation twinning for different materials and testing conditions.

In particular, lower temperatures and/or additional cold rolling (CR), extrusion, or drawing promote the formation of nanotwins. In Fig. 1.14, on a TEM image of atomic resolution of UFG Cu after ECAP and CR at liquid nitrogen temperature, it is possible to clearly observe twins of 10–20 nm in size [39]. It has been also observed that an enhanced strength and ductility can be obtained in nanocrystalline materials due to deformation twinning [39, 40].

Twin GBs (even in nanotwins) typically contain the high-energy facets in addition to the "main" low-energy Σ3 {111}1//{111}2 facet. The observation of such high-energy facets in materials after SPD is an additional instrument for the characterization of non-equilibrium GBs processed by SPD [41].

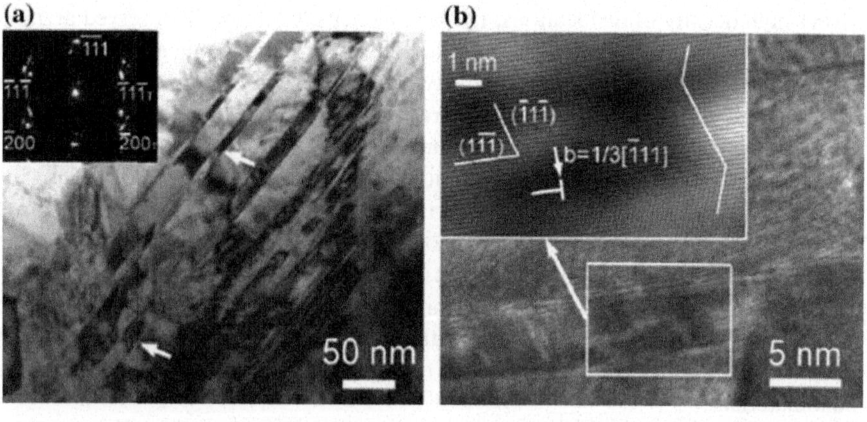

Fig. 1.14 TEM images at a low (**a**) and high (**b**) magnifications of a typical grain with a high density of deformation twins in UFG Cu processed by ECAP with subsequent cold rolling. (Reproduced from [39], Copyright 2006, with permission from John Wiley & Sons)

1.3.3 Segregations in SPD-Processed Nanostructured Alloys

Investigations of the thermal stability as a function of the impurity level in metals as reported in a few cases supply indirect evidence of grain boundary segregation in UFG materials. However, only recently grain boundary segregations have been observed in UFG materials processed by SPD directly. Atom probe tomography (APT) is the main technique used for atomic scale characterization in most of these reports. At the same time, crystallographic information provided by this technique is very limited and usually the grain boundary misorientation cannot be determined. Moreover, there is almost no possibility to make any statistical analysis because of the very small grain boundary areas that are analyzed. Nevertheless, the demonstration of grain boundary segregation in various kinds of SPD-processed metallic materials [22, 37, 42–44, and others] confirms that this is not a marginal feature.

SPD processing resulted in the formation in a UFG AA6061 alloy of the UFG microstructure, which is characterized by Mg, Cu, and Si segregations along planar defects attributed to grain boundaries [42]. The thickness of the layer enriched by solute elements was about 2 nm. The local enrichment was less than 2 at.%. Segregations at grain boundaries were also observed in several Al alloys processed by both ECAP and HPT [43, 44]. The data in Fig. 1.15 [45] show that clusters ~3–5 nm wide are made up by segregations at grain boundaries. Local concentrations up to very high magnitude equal to 25 at.% of Mg in layers of much higher thickness of ~6 ÷ 8 nm were also observed applying APT [44]. It appears, therefore, that among the characteristic features of a non-equilibrium grain boundary with strain-distorted structure, there are larger amounts of segregating atoms with respect to relaxed interfaces and the thickness of the segregated layer may be determined by the distorted layer near the non-equilibrium GB. It follows from APT observations

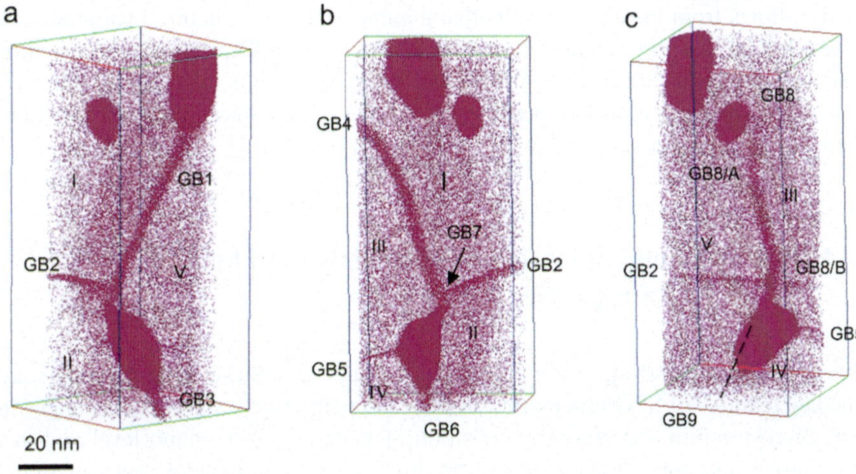

Fig. 1.15 A reconstructed three-dimensional Mg atom map viewed in three directions (**a**), (**b**), and (**c**) to show five grains and nine pieces of GBs of the Al–Zn–Mg–Cu alloy processed by eight-pass ECAP. (Reprinted from [43], Copyright 2011, with permission from Elsevier)

that the distributions of the solute elements may be influenced by the local configurations of the GB, and especially dislocations lying at or near the boundaries.

Thus, the recent results show that GB segregations can exist in different SPD-produced materials, but their behavior is significantly varied depending on the GB structure and the type of alloying elements that in turn depend on the material type and SPD processing regime.

1.3.4 Nanosized Particles and Secondary Phase Precipitations

Solution quenching promotes the formation of particles in many alloys subjected to SPD [46–48, and others]. Many examples of such nano-particles with a size as much as ~10 ÷ 20 nm precipitated in the UFG AA6061 due to ECAP in parallel channels (ECAP-PC) processing is presented in these studies. Nano-particles appeared due to dynamic aging and caused additional precipitation hardening of the alloys. At the same time, it should be noted that parameters of SPD processing usually determine the size and morphology of precipitations. In recent years, the subject of dynamic aging originating from SPD has attracted greater attention because the morphology of precipitates and their formation kinetics differ considerably from the processes of conventional aging and provide for new possibilities in the production of advanced age-hardenable alloys [49, 50].

It has been also observed recently in copper-based alloys that the dynamic aging during high-pressure torsion competes with dynamic dissolution of precipitates [51]. As a result of this competition, a certain steady-state forms between dynamic aging and dynamic dissolution [52]. This is controlled by an accelerated diffusion-like mass transfer between matrix solid-solution and precipitates. As a result of this dynamic equilibrium, a certain steady-state solute concentration c_{ss} is established.

It follows from the above results that a number of nanostructured parameters is characteristic for UFG metals and alloys produced by SPD processing. As will be shown in the following section, these parameters can strongly influence their behavior and, in many demonstrated cases, SPD processing induces unique features in their material properties.

1.4 Nanostructural Design of Materials for Their Advanced Properties

It is apparent that strong grain refinement and unusual phase transformations are the key features of SPD processing, which can result in very significant changes in the microstructure and phase composition of materials at the nano level and this determines changes in their properties both at the mechanical and functional levels. This is the basis for the nanostructural design of materials to improve their properties.

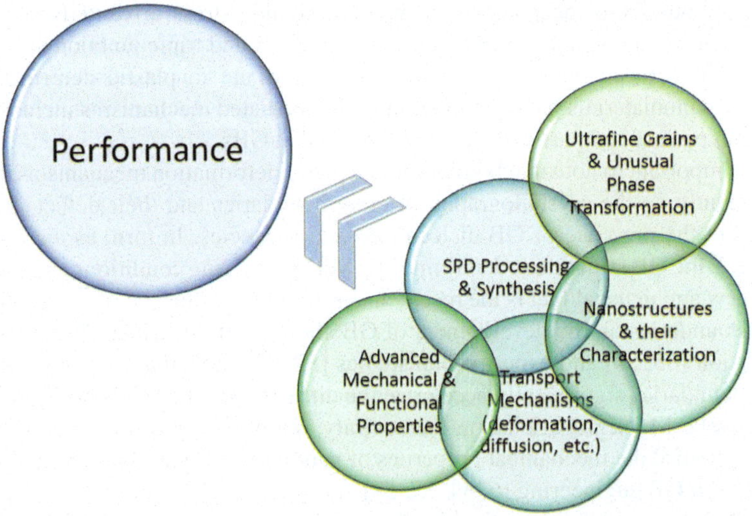

Fig. 1.16 Principles of nanostructural design of bulk nanostructured materials using SPD processing. (Reproduced from [6], Copyright 2014, with permission of John Wiley & Sons)

The concept of this design of materials is schematically illustrated in Fig. 1.16. The scheme modifies and further develops a well-known concept of the contemporary creation of novel materials through the integration of theory and modeling, structure characterization, processing and synthesis, as well as the study of properties. However, nanostructuring of bulk materials deals with a far larger number of structural parameters related to the grain size and shape, the lattice defects in the grain interior, as well as with the grain boundary structure and also the presence of segregations and second-phase nanoparticles. This provides an opportunity to vary the transport mechanisms and therefore can drastically increase the properties.

This is the stage when different materials can be manufactured with unusual mechanical, physical and chemical properties that have not been observed earlier. This new subject is gaining momentum rapidly and, as demonstrated in recent analytical reviews [19, 53, 54], has been the focus of numerous works on achieving very high levels of properties in metallic materials, some polymers and ceramics where such levels are truly unique.

Over the last decade, a number of theoretical and experimental studies has been carried out that convincingly demonstrate changes in the mechanisms of deformation and diffusion processes in nanomaterials, where the above-discussed nanostructural features such as very fine grain sizes, high densities of lattice defects, and grain boundaries affect both the nature and kinetics of these processes.

It is known that deformation mechanisms in nanometals are significantly influenced by grain boundaries, which constitute a large volume fraction in such metals. In general, GB-mediated deformation mechanisms can be divided into two basic categories: (1) "pure" GB mechanisms by the motion of GBs themselves, and (2) GB-mediated mechanisms associated with dislocation emission from GBs [19, 33,

54]. The "pure" GB mechanisms include GB sliding, stress-driven GB migration, grain rotations mediated by GB dislocations, and GB and triple junction diffusional creep. These mechanisms can significantly contribute to plastic deformation in metallic nanomaterials having nanograins. GB-mediated mechanisms include twinning and perfect and partial dislocations emitted at GBs.

It is important to note that both of these types of deformation mechanisms depend significantly on the crystallography of grain boundaries and their defect structure related to the presence of GB dislocations and vacancies. In turn, as discussed in Sect. 1.3, the GB structure is determined by SPD processing conditions and regimes.

A new and unusual fact is the recent discovery of the influence of segregations at grain boundaries on the development of GB sliding and diffusion. Thus, computer simulation [55, 56] and experimental studies [57, 58] show that Zn segregations at grain boundaries significantly accelerate the diffusion kinetics and sliding development, while Mg segregations, on the contrary, inhibit this process. This opens the path to control the mechanical properties of nanomaterials and their thermal stability through GB engineering [6, 34, 37, 58].

In the next sections, we consider the development of nanostructural design of metals and alloys for a number of relevant and exciting problems—superstrength and plasticity, high strength and electrical conductivity, a combination of mechanical and functional properties of metal biomaterials. These problems refer to multifunctional materials, when it is possible to simultaneously improve their properties, which are usually mutually exclusive [19, 46].

Suggested Reading

1. Gleiter H (1989) Nanocrystalline materials. Prog Mater Sci 33:223–315
2. Whang SH (2011) Nanostructured metals and alloys: processing, microstructure, mechanical properties and applications. Woodhead Publishing Limited, Sawston
3. Valiev RZ, Islamgaliev RK, Alexandrov IV (2000) Bulk nanostructured materials from severe plastic deformation. Prog Mater Sci 45:103–189
4. Langdon TG (2013) Twenty-five years of ultrafine-grained materials: achieving exceptional properties through grain refinement. Acta Mater 61:7035–7059
5. Valiev RZ, Zhilyaev AP, Langdon TG (2014) Bulk nanostructured materials: fundamentals and applications. Wiley/TMS, Hoboken

Problems and Tasks

Introduction

1. The yield stress σ_1 of some metal is 200 MPa and the average grain size $d_1 = 1$ μm. Using the Hall–Petch equation, evaluate the average grain size d_2 for the yield stress to double with the friction stress σ_0 set to 20 MPa.
2. List and analyze the reasons why a submicron grain size cannot be achieved using conventional thermomechanical processing methods.

3. Define and analyze the features of referring to materials as "ultrafine-grained materials."
4. Analyze the fundamentals, advantages, and disadvantages of the "bottom-up" approach used to obtain ultrafine-grained materials.
5. Analyze the fundamentals, advantages, and disadvantages of the "top-down" approach used to obtain ultrafine-grained materials.
6. Define the term "severe plastic deformation"; analyze the features of the implementation of the method of severe plastic deformation.
7. List and analyze the reasons for the activation of the processes that accompany the implementation of the severe plastic deformation method and are not characteristic for conventional methods of thermomechanical treatment.

1.1 Principles of SPD and Processing Techniques

1. List and discuss the role of conditions of severe plastic deformation regimes leading to the formation of ultrafine-grained structure.
2. Describe the features of the implementation of the high pressure torsion technique.
3. Using the equation $\varepsilon = \ln\left(\frac{\varphi \cdot r}{h}\right) = \ln\left(\frac{2\pi N \cdot r}{h}\right) = \ln \gamma$

   where φ is the angle of relative rotation of the anvils, r is the distance from the rotation axis passing through the workpiece center, perpendicular to its surface, h is the workpiece thickness and γ is the degree of shear strain estimated using the angle γ, calculate the value of the true logarithmic strain ε: (1) after $N = 1$ revolution; (2) $N = 10$ revolutions, $r = 5$ mm, $h = 1$ mm. Compare the amount of strain obtained by high pressure torsion with the amount that can be achieved using conventional thermomechanical processing methods.
4. Describe the features of the implementation of the equal channel angular pressing technique.
5. Using the equation $\gamma = 2\cot\left(\frac{\Phi}{2} + \frac{\psi}{2}\right) + \psi \operatorname{cosec}\left(\frac{\Phi}{2} + \frac{\psi}{2}\right)$ and $\varepsilon_N = \frac{N\gamma}{\sqrt{3}}$,

   make and analyze the plot of the equivalent strain ε_N vs. the value of internal

   angle Φ at the value of the external angle $\Psi = 20°$ for one pass ($N = 1$) of equal channel angular pressing.
6. Draw and explain possible schemes for realizing equal channel angular pressing routes.
7. Describe the influence of the number of passes and routes on the features of the obtained UFG structure.

1.2 Grain Refinement, Its Mechanisms, and Ultrafine Grains

8. Describe the stages of microstructure evolution during large deformations according to V.V. Rybin.
9. Describe the stages of microstructure evolution during large deformations according to N. Hansen.

10. Describe and analyze the features and formation processes of random dislocation boundaries and geometrically necessary boundaries.

1.3 Nanostructural Architecture of SPD-Processed Materials

11. Grain Boundaries in Nanostructured Materials

    Describe the features of the structure of nonequilibrium grain boundaries in UFG materials obtained by severe plastic deformation techniques.

12. Nanotwins in Metallic Nanomaterials

    Analyze the features of structure and processes of nanotwinning and the role of nanotwins in enhancing the properties of nanomaterials.

13. Segregations in SPD-Processed Nanostructured Alloys

    What are the patterns of segregation formation in SPD-processed nanostructured alloys?

14. Nanosized Particles and Secondary Phase Precipitations

    Describe the features of formation of nanosized particles and secondary phase precipitations in UFG materials.

1.4 Nanostructural Design of Materials for Their Advanced Properties

15. Describe the processes underlying nanostructural design for advanced properties of metals and alloys.

References

1. Gleiter H (1989) Nanocrystalline materials. Prog Mater Sci 33:223–315
2. Whang SH (2011) Nanostructured metals and alloys: processing, microstructure, mechanical properties and applications. Woodhead Publishing Limited, Sawston
3. Valiev RZ, Islamgaliev RK, Alexandrov IV (2000) Bulk nanostructured materials from severe plastic deformation. Prog Mater Sci 45:103–189
4. Langdon TG (2013) Twenty-five years of ultrafine-grained materials: achieving exceptional properties through grain refinement. Acta Mater 61:7035–7059
5. Bridgman PW (1952) Studies in large plastic flow and fracture. McGraw-Hill, New York
6. Valiev RZ, Zhilyaev AP, Langdon TG (2014) Bulk nanostructured materials: fundamentals and applications. Wiley/TMS, Hoboken
7. Smithsonian Institution Archives, Accession 90–105, Science Service Records, Image No. SIA2008-0025
8. Zhilyaev AP, Langdon TG (2008) Using high-pressure torsion for metal processing: Fundamentals and applications. Prog Mater Sci 53:893–979
9. Lowe TC, Valiev RZ (eds) (2000) Proceedings of the NATO ARW on investigations and applications of severe plastic deformation 80, NATO science series. Kluwer Publ, Moscow, 394 p
10. Zhilyaev AP, McNelley TR, Langdon TG (2007) Evolution of microstructure and microtexture in fcc metals during high-pressure torsion. J Mater Sci 42:1517–1528

11. Bondarenko GG, Kabanova TA, Rybalko VV (2020) Fundamentals of materials science. BKL Publishers, Moscow
12. Iwahashi Y, Wang J, Horita Z, Nemoto M, Langdon TG (1996) Principle of equal-channel angular pressing for the processing of ultrafine-grained materials. Scr Mater 35:143–146
13. Nakashima K, Horita Z, Nemoto M, Langdon TG (2000) Development of a multi-pass facility for equal-channel angular pressing to high total strains. Mater Sci Eng A 281:82–87
14. Cao Y, Ni S, Liao X, Song M, Zhu Y (2018) Structural evolutions of metallic materials processed by severe plastic deformation. Mater Sci Eng R 133:1–59
15. Romanov AE (2004) Importance of disclinations in severe plastically deformed materials. In: Zehetbauer MJ, Valiev RZ (eds) Nanomaterials by severe plastic deformation. WILEY-VCH, Weinheim
16. Rybin VV (1986) Large plastic deformations and fracture of metals. Metallurgiya, Moscow
17. Hughes DA, Hansen N (2000) Microstructure and strength of nickel at large strains. Acta Mater 48:2985
18. Hughes DA, Hansen N (1997) High angle boundaries formed by grain subdivision mechanisms. Acta Mater 45:3871
19. Valiev RZ, Estrin Y, Horita Z, Langdon TG, Zehetbauer MJ, Zhu YT (2016) Fundamentals of superior properties in bulk nanoSPD materials. Mater Res Lett 4:1–21
20. Straumal BB, Kilmametov AR, Mazilkin AA, Kogtenkova OA, Kurmanaeva L, Korneva A, Zięba P, Baretzky B (2015) Phase transitions induced by severe plastic deformation: steady-state and equifinality. Int J Mater Res 106:657–664
21. Figueiredo R, Langdon TG (2021) Deformation mechanisms in ultrafine-grained metals with an emphasis on the Hall-Petch relationship and strain rate sensitivity. J Mater Res Tech 14:137–159
22. Sauvage X, Wilde G, Divinski SV, Horita Z, Valiev RZ (2012) Grain boundaries in ultrafine grained materials processed by severe plastic deformation and relared phenomena. Mater Sci Eng A 1:540
23. Straumal BB, Kilmametov AR, López GA, López-Ferreño I, Nó ML, San Juan J, Hahn H, Baretzky B (2017) High-pressure torsion driven phase transformations in Cu–Al–Ni shape memory alloys. Acta Mater 125:274–285
24. Kilmametov A, Straumal BB, Mazilkin AA, Gornakova AS, Kriegel MJ, Fabrichnaya OB, Rafaja D, Hahn H (2017) Transformations of α' martensite in Ti–Fe alloys under high pressure torsion. Scr Mater 136:46–49
25. Kilmametov A, Mazilkin AA, Straumal BB, Gornakova AS, Fabrichnaya OB, Kriegel MJ, Rafaja D, Hahn H (2018) The $\alpha \rightarrow \omega$ and $\beta \rightarrow \omega$ phase transformations in Ti–Fe alloys under high-pressure torsion. Acta Mater 144:337
26. Gleiter H (2000) Nanostructured materials: basic concepts and microstructure. Acta Mater 48:1
27. Edalati K, Horita Z (2016) A review on high-pressure torsion (HPT) from 1935 to 1988. Mater Sci Eng A 652:325–352
28. Mckenzie PWJ, Lapovok R (2010) ECAP with back pressure for optimum strength and ductility in aluminium alloy 6016. Part 1: microstructure. Acta Mater 58(9):3198–3211
29. Rank O, Pippan R (2019) Saturation of grain refinement during severe plastic deformation of single phase materials: reconsiderations, current status and open questions. Mater Trans 60:1270
30. Zehetbauer MJ, Zhu YT (eds) (2009) Bulk nanostructured materials. Wiley, Weinheim
31. Kolobov YR, Grabovetskaya GP, Ivanov KV, Ivanov MB (2002) Grain boundary diffusion and mechanisms of creep of nanostructured metals. Interface Sci 10:31
32. Divinski SV, Reglitz G, Rösner H, Wilde G, Estrin Y (2011) Self-diffusion in Ni prepared by severe plastic deformation: effect of non-equilibrium grain boundary state. Acta Mater 59:1974
33. Wilde G, Divinski S (2019) Grain boundaries and diffusion phenomena in severely deformed materials. Mater Trans 60:1302

34. Raabe D, Herbig M, Sandlöbes S, Li Y, Tytko D, Kuzmina M, Ponge D, Choi PP (2014) Grain boundary segregation engineering in metallic alloys: a pathway to the design of interfaces. Curr Opin Solid State Mater Sci 18:253–261
35. Horita Z, Smith DJ, Furukawa M, Nemoto M, Valiev RZ, Langdon TG (1996) An investigation of grain boundaries in submicrometer-grained Al-Mg solid solution alloys using high-resolution electron microscopy. J Mater Res 11:1880
36. Wang J, Horita Z, Furukawa M, Nemoto M, Tsenev NK, Valiev RZ, Ma Y, Langdon TG (1993) An investigation of ductility and microstructural evolution in an Al-3% Mg alloy with submicron grain size. J Mater Res 8:2810–2818
37. Schuh A, Lu K (2021) Stability of nanocrystalline metals: the role of grain-boundary chemistry and structure. MRS Bull 46:225–235
38. Zhu YT, Liao XZ, Wu XL (2012) Deformation twinning in nanocrystalline materials. Prog Mater Sci 57:1–62
39. Zhao Y, Bingert JF, Liao X, Cui B, Han K, Sergueeva AV, Mukherjee AK, Valiev RZ, Langdon TG, Zhu YT (2006) Simultaneously increasing the ductility and strength of ultra-fine-grained pure copper. Adv Mater 18:2949
40. Lu K, Lu L, Suresh S (2009) Strengthening materials by engineering coherent internal boundaries at the nanoscale. Science 324:349–352
41. Straumal BB, Kogtenkova OA, Gornakova AS, Sursaeva VG, Baretzky B (2016) Review: grain boundary faceting-roughening phenomena. J Mater Sci 51:382–404
42. Nurislamova GV, Sauvage X, Murashkin MY, Islamgaliev RK, Valiev RZ (2008) Nanostructure and related mechanical properties of an Al-Mg-Si alloy processed by severe plastic deformation. Philos Mag Lett 88:459
43. Sha G, Yao L, Liao X, Ringer SP, Duan ZC, Langdon TG (2011) Segregation of solute elements at grain boundaries in an ultrafine grained Al–Zn–Mg–Cu alloy. Ultramicroscopy 111:500–505
44. Valiev RZ, Enikeev NA, Murashkin MY, Kazykhanov VU, Sauvage X (2010) On the origin of extremely high strength of ultrafine-grained Al alloys produced by severe plastic deformation. Scr Mater 63:949
45. Sha G, Ringer SP, Duan ZC, Langdon TG (2009) An atom probe characterisation of grain boundaries in an aluminium alloy processed by equal-channel angular pressing. Int J Mater Res 100:1674–1678
46. Sabirov I, Enikeev NA, Murashkin MY, Valiev RZ (2015) Bulk nanostructured materials with multifunctional properties, SpringerBriefs in materials. Springer, p 64
47. Valiev RZ, Murashkin MY, Bobruk EV, Raab GI (2009) Grain refinement and mechanical behavior of the Al alloy, subjected to the new SPD technique. Mater Trans 50:87–91
48. Cheng S, Zhao YH, Zhu YT, Ma E (2007) Optimizing the strength and ductility of fine structured 2024 Al alloy by nano-precipitation. Acta Mater 55:5822–5832
49. Valiev RZ, Murashkin MY, Sabirov I (2014) A nanostructural design to produce high strength Al alloys with enhanced electrical conductivity. Scr Mater 76:13–16
50. Ma K, Zheng Y, Dasari S, Zhang D, Fraser HL, Banerjee R (2021) Precipitation in nanostructured alloys: a brief review. MRS Bull 46:250–257
51. Straumal BB, Kilmametov AR, Korneva A, Mazilkin AA, Straumal PB, Zięba P, Baretzky B (2017) Phase transitions in Cu-based alloys under high pressure torsion. J Alloys Compd 707:20–26
52. Straumal BB, Pontikis V, Kilmametov AR, Mazilkin AA, Dobatkin SV, Baretzky B (2017) Competition between precipitation and dissolution in Cu–Ag alloys under high pressure torsion. Acta Mater 122:60–71
53. Horita Z, Edalati K (2020) Severe plastic deformation for nanostructure controls. Mater Trans 61:2241
54. Ovid'ko IA, Valiev RZ, Zhu YT (2018) Review on superior strength and enhanced ductility of metallic nanomaterials. Prog Mater Sci 94:462–540

55. Suresh S (ed) (2000) The millennium special issue. A selection of major topics in materials science and engineering: current status and future directions. Acta Mater 48:1–384
56. Petrik MV, Kuznetsov AR, Enikeev NA, Gornostyrev YN, Valiev RZ (2018) Peculiarities of interactions of alloying elements with grain boundaries and the formation of segregations in Al-Mg and Al-Zn alloys. Phys Met Metallogr 119:607–612
57. Edalati K, Horita Z, Valiev RZ (2018) Transition from poor ductility to room-temperature superplasticity in a nanostructured aluminum alloy. Sci Rep 8:6740
58. Chinh NQ, Murashkin MY, Bobruk EV, Labar J, Gubicza J, Kovacs Z, Ahmed AQ, Maier-Kiener V, Valiev RZ (2021) Ultralow-temperature superplasticity and its novel mechanism in ultrafine-grained Al alloys. Mater Res Lett 9:475–482

Chapter 2
Mechanical Properties of UFG Metals

Mechanical properties relate to the key properties for metallic materials. However, the level of mechanical properties depends on a number of factors such as a type of crystal lattice, stacking fault energy as well as degree, strain rate and temperature, initial microstructure, etc.

As for UFG metallic materials, other conditions being equal, the specific microstructure resulting from SPD processing is of high importance. The developed network of high-angle grain boundaries provides an opportunity to achieve very high strength, manifestation of low-temperature and high-speed superplasticity, achievement of elevated fatigue properties, etc.

Mechanical properties of UFG metals and alloys and their nature have important features in comparison with the properties of their coarse-grained counterparts at low as well as at high temperatures.

2.1 Mechanical Properties at Ambient Temperature

Let us consider the mechanical properties of UFG metallic materials from monotonic to cyclic deformation sequence.

2.1.1 Superstrength and Hardening Mechanisms

A primary advantage of UFG materials from SPD processing is strength enhancement. As noted above, it is usually well described by the Hall–Petch relationship (see Eq. 1.1), and the enhancement of yield strength with grain size reduction in compliance with this relation was experimentally observed in many studies, but for nanosized grains (20–50 nm), this relation is reported to be typically violated so that

R. Z. Valiev et al., *Ultrafine-Grained Materials*, The Minerals, Metals & Materials Series, https://doi.org/10.1007/978-3-031-31729-3_2

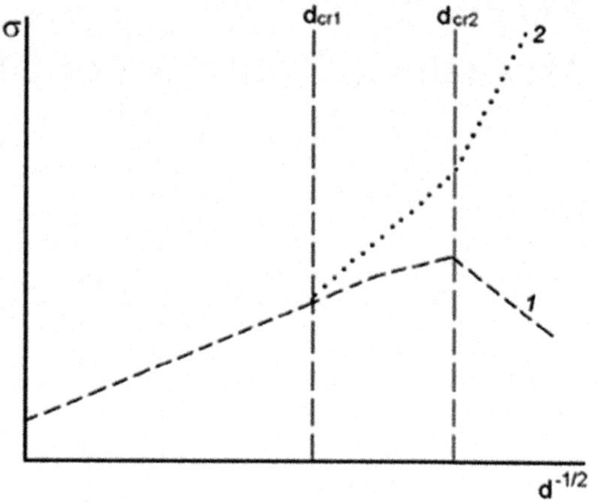

Fig. 2.1 The two types of the Hall-Petch slopes within different characteristic length scales. (Reproduced from [4], Copyright 2011, with permission of Kovove Materialy-Metallic Materials)

the Hall–Petch plot deviates from linear dependence at lower stress values and its slope k_y often becomes negative (the curve 1 in Fig. 2.1). In recent years, this problem has been widely analyzed in both experimental and theoretical studies [1–3]. At the same time, the Hall–Petch relationship breakdown is not observed in ultrafine-grained materials with a mean grain size of 100–1000 nm typically produced by SPD processing. Moreover, recently it was shown that UFG pure metals and especially alloys can exhibit a considerably higher strength than the Hall–Petch relationship predicts for the range of ultrafine grains (the curve 2 in Fig. 2.1) [1, 4, 5]. The nature of such superstrength was recently studied in detail for the case of super-strong UFG Al alloys, stainless steels, and other alloys produced by severe plastic deformation.

For example, in the Al alloys 1570 and 7475, homogeneous UFG structures, with a grain size of about 100 nm, were formed as a result of HPT, as shown in Fig. 2.2. It was also determined that HPT processing has an observable effect on the value of the crystal lattice parameter a of Al alloys. In particular, in the AA1570, its value due to straining was reduced considerably in comparison with the initial state, from 4.0765 ± 0.0001 Å to 4.0692 ± 0.0003 Å, which results from the formation of Mg segregations at grain boundaries [6].

Figure 2.3 [6] illustrates the data for a number of Al alloys presented in the form of the Hall–Petch relation in which the yield stress ($\sigma_{0.2}$) is plotted against the inverse square root of the grain size ($d^{-1/2}$) for a UFG AA1100 produced by accumulative roll bonding(ARB)-rolling and subsequent heat treatment as well as for an ECAP-processed Al–3% Mg alloy. For the Hall–Petch relationship in the AA1100, the following parameters were set: $\sigma_0 = 6.0$ MPa and $k_y = 105$ (for the grain sizes in μm); for the ECAP-processed Al–3% Mg alloy, $\sigma_0 = 62$ MPa and $k_y = 149$. Figure 2.3 also shows the data obtained for the coarse-grained and UFG AA1570 and AA7475 alloys.

From the available data, it is seen that the σ_y values for the coarse-grained solution-treated alloys are close to the results for the Al–3% Mg alloy. However, for the UFG states in the 1570 and 7475 alloys with a grain size of 100–130 nm, the

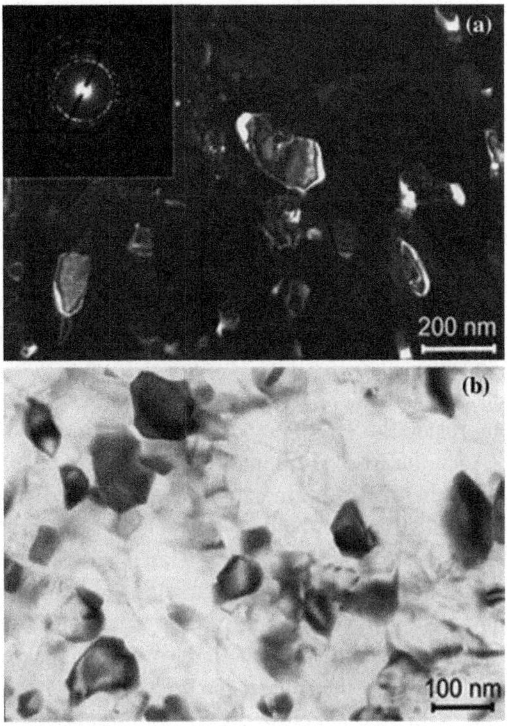

Fig. 2.2 A typical TEM dark-field image of the UFG 1570 alloy with a corresponding SAED (**a**), a bright-field image (**b**). (Reprinted from [6], Copyright 2011, with permission from Elsevier)

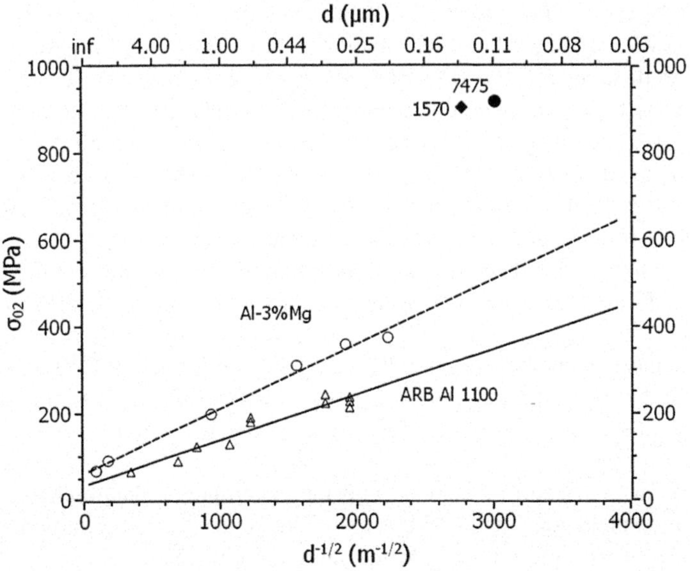

Fig. 2.3 The Hall–Petch relation for the Al alloys 1100 and Al–3 % Mg and data on the yield stresses of the UFG alloys 1570 and 7475. (Reprinted from [6], Copyright 2011, with permission from Elsevier)

value σ_y is considerably higher than that calculated from the Hall–Petch relationship.

The Hall–Petch plots for UFG commercially pure (CP) Ti and Armco-Fe have shown similar tendencies.

Thus, the new phenomena of "superstrength" of nanostructured metals and alloys produced by SPD can be observed for various materials reflecting the "positive" slope of the curve corresponding to the Hall–Petch relationship as demonstrated in Fig. 2.1 (curve 2). "Superstrength" in bulk nanostructured materials can be of different origin. First, it can result from several well-known strengthening mechanisms, i.e., solid solution strengthening, precipitation strengthening, and dislocation strengthening [2, 3, 6]. However, the influence of new strengthening mechanisms is also possible. This arises from the change in grain boundary structure and leads to further strength enhancement [4, 7, 9]. Let us consider this point in more detail.

Similar to experimental studies of nanomaterials obtained by vacuum deposition [5], where, for the first time, was reported a "positive" slope of the curve corresponding to the Hall–Petch relationship (Fig. 2.1), for the grain size range $d_{cr2} < d < d_{cr1}$ and for $d < d_{cr1}$, related to the presence of grain boundary segregations, the presence of these two characteristic grain size values is also valid for nanostructured SPD materials. Indeed, as it has been already pointed out, the presence of non-equilibrium grain boundaries is typical for the majority of SPD-processed materials, but their influence on mechanical properties becomes considerable when the grain size is below 500–1000 nm [1]. When the grain size decreases down to 100 nm and less, grain boundary segregations provide a significant contribution to the overall strength. For example, it was shown that high strength of the UFG Al alloys was directly related to the formation of Mg segregations at grain boundaries revealed in the alloys by the 3D APT technique (Fig. 1.15) [6, 8]. Because, in UFG materials, deformation takes place by dislocations generated at grain boundaries and moving through a grain to be captured by an opposite grain boundary, the rate-controlling mechanism is the "dislocation–grain boundary" interaction. In this case, the elevated concentration of solutes in grain boundaries can suppress emission of dislocations from such boundaries due to solute drag, and as a result, the stress needed to emit a dislocation increases. Recently, this finding was confirmed for the case of UFG stainless steel and some other alloys produced by SPD at different temperatures [8, 9].

These findings on the role of SPD-induced grain boundary (GB) segregations open a way for a new direction in the nanostructural design of SPD-processed materials and also may be used for grain boundary engineering of various nanomaterials.

This point is in good agreement with the experimental data and also makes it possible to explain considerably higher strength values observed in the UFG alloys in comparison with pure metals. At the same time, there is the task for oncoming research to explore the segregations and determine which elements, and the level of their concentrations, produce the most influence on strength enhancement in bulk nanostructured materials. This point is also important in answering the question of the precise origin of very high strength in nanostructured multiphase alloys that possess the most strength among metallic nanomaterials, which was recently discussed [10].

2.1.2 Ductility and Strategies for Its Improvement

Ductility in metallic materials, which is measured under tensile deformation mode, is another very important property for their structural use. It is essential for metal-forming operations as well as to avoid catastrophic failure in load-bearing applications during their service life. Two measures of ductility typically can be considered [11]. First, uniform elongation is the maximum strain where homogeneous plastic flow (uniform reduction of cross-sectional area) is still observed and beyond which a necking begins. The uniform elongation is determined by competition between plastic flow and materials resistivity to macro-localization of plastic flow (i.e., strain hardening). It can be important in stretch-metal-forming operations. The second measure, elongation to failure (often referred to as total elongation to fracture or engineering strain at fracture), is the sum of the uniform elongation and the necking elongation. The latter is controlled by competition between localized plastic flow and fracture processes. The uniform elongation is not affected much by the sample geometry and size, and therefore is more suitable for use as a measure of ductility, especially for the study of nanostructured materials where small samples with various geometries are typically used.

It has been a long-standing goal for materials scientists to synthesize structural materials with balanced combinations of high strength and high ductility. As mentioned above, during the last decade, it has been widely demonstrated that a major grain refinement, down to the nanometer range, may lead to a very high hardness and strength in various metals and alloys, but nevertheless, these materials typically exhibit low ductility [11, 12]. A similar tendency is well known for metals subjected to strengthening by other processes such as rolling, extrusion, or drawing. Strength and ductility are the key mechanical properties of any material, but these properties typically have opposing characteristics. Thus, materials may be strong or ductile, but they are rarely both. The reason for this dichotomy is of a fundamental nature. Indeed, the strain hardening mechanisms associated with the accumulation of dislocations may not be effective in ultrafine grains and the low strain hardening leads to low ductility, as discussed below.

Typically, in UFG metallic materials, deformation localization (necking) occurs at the very early stage of plastic deformation (1–3%), followed by a specimen failure. Various strategies to improve low ductility of these metals and alloys have been proposed [1, 11–13]. These strategies are divided into two groups which one can define as (1) "mechanical" strategies and (2) "microstructural" strategies. The "mechanical" strategies employ the mechanical characteristics of the UFG materials such as their work hardening ability and/or strain rate sensitivity. These mechanical characteristics can be varied by changing the testing parameters such as temperature and/or strain rate. The "microstructural" strategies are based on the approach of nanostructural design.

For tensile behavior, the uniform strain is usually in good agreement with the well-known Considère criterion,

$$\left(\frac{d\sigma}{d\varepsilon}\right) \geq \sigma \tag{2.1}$$

which is a geometric criterion stating that when the work hardening rate, $(d\sigma/d\varepsilon)$, decreases to the level of the flow stress, σ, macro-localization of plastic deformation (necking) should occur, resulting in a specimen failure. The UFG metallic materials are often characterized also by increased strain rate sensitivity even at low temperatures compared to their coarse-grained counterparts. Therefore, the Hart criterion [11] is more appropriate to predict the uniform elongation of these metals and alloys:

$$\left(\frac{d\sigma}{d\varepsilon}\right) = (1-m)\sigma \tag{2.2}$$

where m is the strain rate sensitivity. It follows from Eq. 2.2 that the uniform elongation can be enhanced by (1) increasing the work hardening rate $d\sigma/d\varepsilon$ and/or (2) the strain rate sensitivity m.

Meanwhile, since the mechanical characteristics of the materials are also determined by their microstructure, it can be concluded that manipulation with the microstructure is the major tool to enhance the ductility of UFG alloys.

In this connection, it is important to consider with more details the three different approaches that have been already proposed in earlier studies on UFG metals produced by SPD.

In the first study, high-purity (99.996%) Cu was processed at room temperature using ECAP with a 90° clockwise rotation around the billet axis between consecutive passes in route B_C [14]. The strength and ductility were measured by uniaxial tensile tests, and the resulting engineering stress–strain curves are shown in Fig. 2.4 for the Cu samples tested at room temperature in the initial coarse-grained condition and in three processed states. It is apparent that the initial coarse-grained Cu with a grain size of about 30 μm has a very low yield stress with significant strain hardening and a large elongation to failure. Cold rolling of the Cu with reduction ratio of 60% significantly increases the strength (curve 2 in Fig. 2.4), but dramatically decreases the elongation to failure. This result is consistent with the classical mechanical behavior of metals that are plastically deformed.

The same tendency is true also for Cu subjected to two passes of ECAP. However, further straining of Cu to 16 passes of ECAP (curve 4 in Fig. 2.4) simultaneously increases both the strength and ductility. Furthermore, the increase in ductility is much more significant than the relatively minor increase in strength. Thus, the data shown in Fig. 2.4 for ECAP-processed Cu clearly demonstrate an enhancement of strength as well as ductility with accumulated deformation due to an increase in the

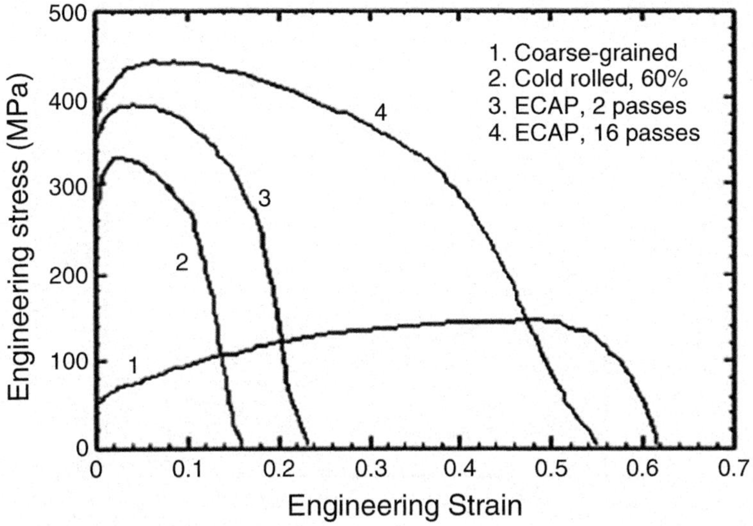

Fig. 2.4 Tensile engineering stress–strain curves for Cu tested at 22 °C with a strain rate of 10^{-3} s^{-1}: The processing conditions for each curve are indicated. (Reprinted by permission from Springer Nature [14], Copyright 2002)

number of passes from 2 to 16. This is a very remarkable result that, for the first time, was observed in metals processed by plastic deformation. Accordingly, this effect was termed the "paradox of strength and ductility in SPD-processed UFG metals," and the principles of this paradox are illustrated in Fig. 2.5, where it is apparent that conventional metals lie within the lower shaded quadrant [14]. As shown in Fig. 2.5, for Cu and Al, cold rolling (the reduction in thickness is marked by each datum point) increases the yield strength but decreases the elongation to failure. The extraordinary combination of high strength and high ductility, shown in Fig. 2.5 for the nanostructured Cu and Ti after SPD processing, clearly sets them apart from the other coarse-grained metals.

In recent years, similar tendencies have been reported in a number of UFG metals (see [11, 15]), including Al, Cu, Ni, and Ti produced by SPD processing via ECAP, HPT, and ARB. The origin of this phenomenon has been proposed to associate with an increase in the fraction of high-angle grain boundaries with strain enhancement and with a consequent change in the dominant deformation mechanisms due to the increasing tendency for the occurrence of grain boundary sliding and grain rotation. Implementation of these mechanisms is confirmed by higher strain rate sensitivity of UFG materials that manages plasticity, following Eq. 2.2. A high value for strain rate sensitivity indicates viscous flow and renders the material more resistant to necking and, therefore, more ductile. It was recently demonstrated that the *m*-value in the UFG Al alloys at room temperature can be increased up to 0.24 via manipulation with the chemical composition of grain boundaries [16] that can lead to the extraordinarily high ductility at room temperature.

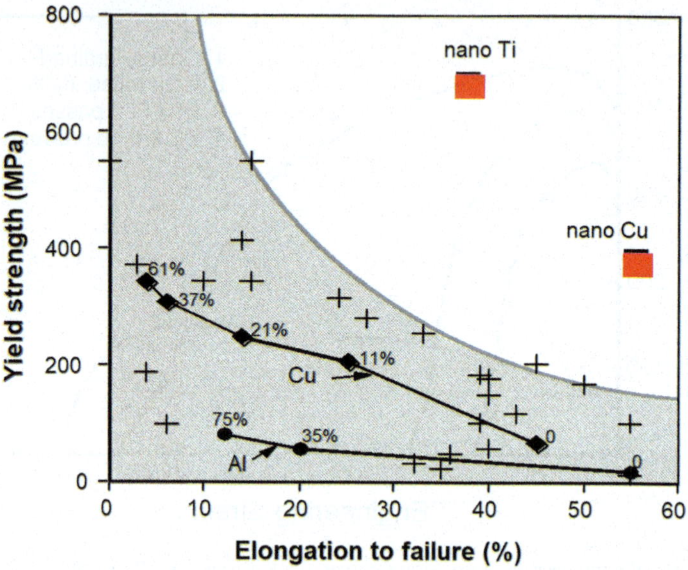

Fig. 2.5 The paradox of strength and ductility in metals subjected to SPD: the extraordinary combination of high strength and high ductility in UFG Cu and Ti processed by SPD (two upper points) clearly sets them apart from conventional coarse-grained metals (lower points relating to metals of 99.5–99.9% purity). (Reprinted by permission from Springer Nature [14], Copyright 2014)

Another approach to enhance ductility is based on introduction of a bimodal distribution of grain sizes [17]. In this work UFG Cu was produced through a combination of ECAP and subsequent rolling at the liquid nitrogen temperature prior to heating to ~450 K. This processing route resulted in the formation of a bimodal structure of micrometer-sized grains, with a volume fraction of around 25%, embedded in a matrix of UFG grains. The material exhibited an extraordinarily high ductility but also retained a very high strength. The reason for this behavior is that while the UFG structure provides strength, the embedded larger grains stabilize the tensile deformation of the material. Other evidences for the importance of grain size distribution come from investigations of Zn, Cu, and Al alloys [13]. Furthermore, the investigation of Cu showed that bimodal structures may increase the ductility not only during tensile testing but also during cyclic deformation. This observation is also important in improving the fatigue properties of materials.

The third strategy is associated with the microstructural modification that may improve ductility and strength due to modifying the nanostructural features (precipitations, nanoclusters, etc.) to promote accumulation of dislocations [18]. This idea was experimentally realized particularly on Al alloys subjected to ECAP and subsequent artificial aging to precipitation within the grains of nanoparticles [19]. Introduction of artificial aging after ECAP has an important influence on the stress–strain behavior at room temperature, as demonstrated in Fig. 2.6, where the tensile stress–strain curves are shown after ECAP and after solution treatment (ST), CR,

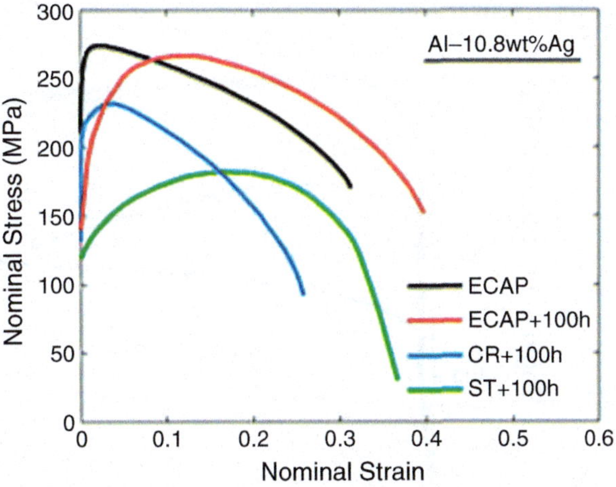

Fig. 2.6 Tensile plots of stress versus strain at room temperature for the Al–10.8% Ag alloy after solution treatment (ST) or cold-rolling (CR) with aging at 373 K for 100 h or ECAP without subsequent aging and ECAP with aging at 373 K for 100 h. (Reproduced from [19] Copyright 2005, with permission of John Wiley & Sons)

and ECAP with additional aging for 100 h at 373 K. Thus, ST and artificial aging give a reasonable tensile strength, an extensive region of uniform strain and good ductility, whereas CR and aging give an increased strength, but very limited uniform strain and a marked reduction in the total ductility. For the ECAP condition, the strength is high in the absence of aging, but there is a negligible region of uniform strain and no significant strain hardening. By contrast, the sample processed by ECAP and aged for 100 h shows a similar high strength, a region of strain hardening, and good ductility. These and other similar results demonstrate the potential for producing high strength and good ductility in precipitation-hardened alloys.

The next way in nanostructural design to improve strength and ductility relies on introduction of high density of twins, which could also assist effective dislocation accumulation inside grains. In SPD-processed materials, nanotwins can be introduced via low-temperature processing and/or deformation at high strain rate [11]. For example, nanostructured Cu with twins was produced by ECAP followed by cryogenic drawing and cryogenic rolling [20]. Dynamic plastic deformation (deformation at high strain rate) in combination with liquid nitrogen temperature also has allowed increasing dramatically the strength of Cu retaining ductility.

It is worth also noting that in the nanostructured metals processed by SPD, both strength and ductility are increased when mechanical tests are conducted at lower temperature and/or high strain rate. As an example, Fig. 2.7 displays the tensile engineering stress–strain curves of UFG Ti with a grain size of 260 nm tested at room temperature and 77 K [21]. At room temperature, the Ti has some ductility and a small uniform elongation, as shown by curve A obtained at a strain rate of 10^{-3} s^{-1}. However, at 77 K, the strength of the material is drastically elevated to ~1.4 GPa. There is also a simultaneous increase in the elongation to failure, and this increases with strain rate up to a maximum close to ~20%, as shown in Fig. 2.7 where curves B–D are for strain rates of 10^{-3} s^{-1}, 10^{-2} s^{-1}, and 10^{-1} s^{-1}, respectively. These results for strength and ductility are better than, or at least comparable to, those of Ti alloys with a high percentage of alloying elements. Here, pronounced necking is delayed

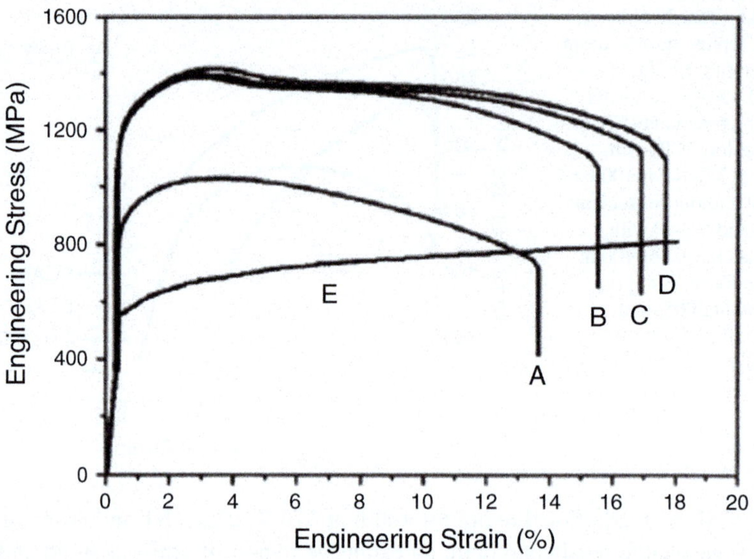

Fig. 2.7 Engineering stress–strain curves for nanostructured Ti where curve A is for testing at room temperature at a strain rate of 1×10^{-3} s $^{-1}$ and curves B–D for the same Ti tested at 77 K for strain rates of 1×10^{-3} s $^{-1}$, 1×10^{-2} s $^{-1}$ and 1×10^{-1}. (Reproduced from [21], Copyright 2004, with permission of John Wiley & Sons)

even for this very strong metal, resulting in a large area under the stress–strain curve and a generally tough behavior of the material.

In conclusion, recent results show that grain refinement by SPD can lead to a unique combination of strength and ductility in metallic materials. Such superior mechanical properties are highly desirable in the development of advanced structural nanomaterials for the next generation applications. However, the achievement of these properties is associated with the tailoring of specific microstructures which, in turn, are determined by the precise processing regimes and the nature of any further treatments. This strategy represents the art and science of nanostructuring for advanced properties using SPD techniques [11, 22].

2.1.3 *Fatigue Behavior*

During their service, modern machines and mechanisms are subjected to cyclic deformations caused by stresses of variable magnitude and character. The value of these stresses does not exceed ultimate tensile strength and does not lead to an instant fracture of the material subjected to cyclic loading. However, there is a gradual degradation of the microstructure and the strength properties of the material decrease to some extent. This cyclic loading behavior of the material is called fatigue behavior [23].

To investigate fatigue behavior, the material is subjected to alternating loads with a fixed amplitude of stress or to cyclic loading with a fixed amplitude of deformation. The number of cycles to failure N_f (fatigue life) is determined for a given range of stresses $\Delta\sigma$ or range of strain $\Delta\varepsilon_t$.

Fatigue occurs when a material is subjected to repeated loading and unloading. If the loads are above a certain threshold, microscopic cracks will begin to form at the stress concentrators such as the surface, persistent slip bands (PSBs), and microstructural heterogeneities (second-phase precipitates, particles, pores, etc.). Eventually, a crack will reach a critical size, the crack will propagate suddenly, and the structure will fracture. The nominal maximum stress values that cause such failure may be much lower than the ultimate tensile strength or yield strength of material. There are two main forms of fatigue: high-cycle fatigue (HCF), and low-cycle fatigue (LCF). In this chapter, we focus on HCF and LCF of ultrafine-grained metallic materials. The principal distinction between HCF and LCF is the region of the stress–strain curve, where the repetitive application of load (and resultant deformation or strain) is taking place. HCF is characterized by low-amplitude high-frequency elastic strains, whereas LCF is the mode of material degradation when plastic strains are induced [23].

The fatigue properties of UFG metallic materials have been widely studied over the last two decades, and comprehensive overviews of these activities can be found [24, 25]. Let us consider fatigue behavior of UFG materials for HCF and LCF regimes.

Although grain refinement down to ultrafine or nanoscale can dramatically enhance tensile strength of metallic materials (see Sect. 2.1.1), their HCF properties are often not enhanced as significantly as tensile strength. For instance, the ratio of the endurance fatigue limit σ_{fo} to the ultimate tensile strength σ_{UTS} (σ_{fo}/σ_{UTS}) was found to be lower for most of the UFG metallic materials compared to their coarse-grained counterparts [24, 25].

Such a little improvement (or no improvement) in fatigue life usually observed in the UFG metallic materials in HCF regime can be related to insignificant increase in resistance against crack nucleation in these materials, since the fatigue life in the HCF is determined by the resistance of material to the crack nucleation. As it is difficult for slip bands to pass through the HAGBs, intergranular cracking by piling up of dislocations at grain boundaries is more likely to occur at HAGBs than in grain interiors or at LAGBs, where dislocations can be transported into the adjacent grain by shear bands.

The HCF response in UFG metallic materials is also strongly determined by their chemical composition. For example, it was shown that increasing Mg content in the Al alloys resulted in longer fatigue life for the same stress amplitude [25]. Significant improvement of fatigue performance was reached via nanostructuring also in Ti-based materials. For example, in CP Ti, grain refinement via ECAP led to an increase in the fatigue limit by nearly 50% [26]. It can also be further enhanced, if the SPD-processed Ti is subjected to additional TMT and annealing, as shown in Fig. 2.8 [26]. Similar effect was also observed in the SPD-processed Ti–6Al–4 V ELI alloy [27].

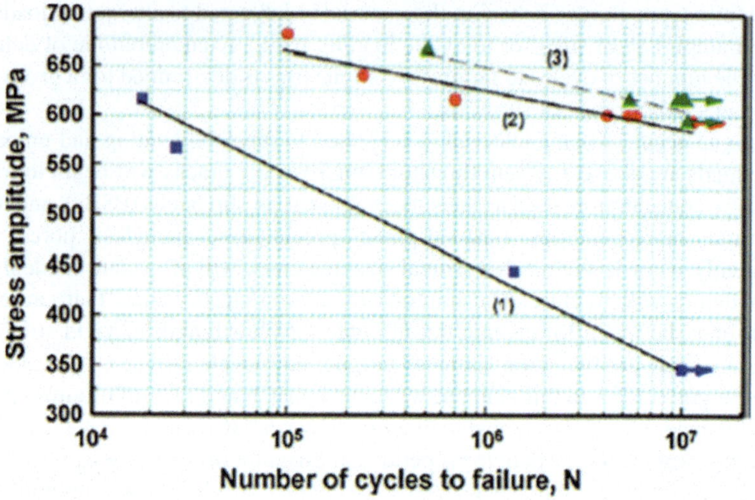

Fig. 2.8 Dependence of the stress amplitude on the number of cycles for smooth samples of coarse-grained CP Ti (curve 1), nanostructured CP Ti after ECAP and thermal mechanical treatment (TMT) (curve 2), and after ECAP + TMT and additional annealing at 350 °C for 6 h (curve 3). (Reprinted from [26], Copyright 2009, with permission from Elsevier)

In the very high-cycle fatigue (VHCF) regime, where the number of cycles to failure is much higher than 10^7, the fatigue life of the UFG pure metals remains high compared to the coarse-grained ones, as was observed in pure Cu and pure Al [28]. However, the VHCF behavior of the nanostructured alloys can be influenced by their microstructure, since the strength of the weakest phase and the area of maximum localized plastic deformation determine the crack initiation sites in the VHCF regime [28].

The LCF behavior of the nanostructured metallic materials is very complex, since ultrafine and nanograins have a low ability to sustain cyclic loads in the LCF regime. This was related to the limited ductility on monotonic and cycling deformation promoting early crack initiation as well as higher fraction of grain boundaries favorable for crack propagation in the nanomaterials [25]. Unlike coarse-grained metals, their nanostructured counterparts usually show no significant increase of stress amplitude (cyclic hardening) (Fig. 2.9) or its decrease (cyclic softening) when they are cyclically deformed under constant strain amplitude [29]. The cyclic softening ratio in this case is nearly zero. This kind of response strongly depends on the chemical composition and microstructure of metals and alloys. The nanostructure in the SPD-processed pure metals is usually unstable, both mechanically and thermally. Microstructural instabilities in the form of dynamic grain coarsening in the nanostructured pure metals (Fig. 2.10) and/or formation of fatigue shear bands have been proposed (Fig. 2.10) as possible reasons for softening effect [29, 30].

For instance, UFG pure Al showed cyclic softening [25], whereas an increase of Mg content in the Al alloys suppressed cyclic softening and even increased cyclic

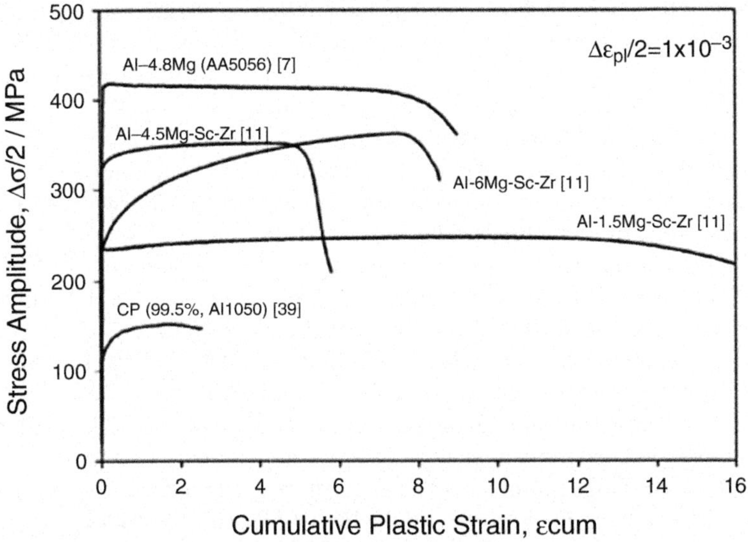

Fig. 2.9 Cyclic response of ECAP-processed Al and various Al–Mg alloys under the same constant plastic strain amplitude. (Reprinted from [25], Copyright 2011, with permission from Elsevier)

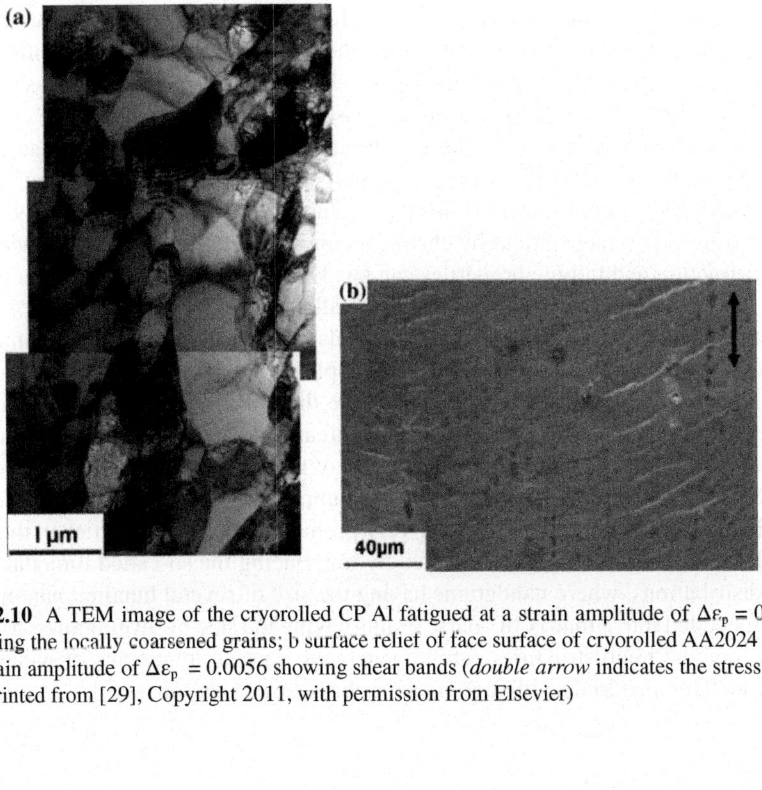

Fig. 2.10 A TEM image of the cryorolled CP Al fatigued at a strain amplitude of $\Delta\varepsilon_p = 0.0015$ showing the locally coarsened grains; b surface relief of face surface of cryorolled AA2024 tested at strain amplitude of $\Delta\varepsilon_p = 0.0056$ showing shear bands (*double arrow* indicates the stress axis). (Reprinted from [29], Copyright 2011, with permission from Elsevier)

hardening effect. The LCF life is mainly controlled by fatigue crack propagation. The fatigue crack growth behavior in the UFG Al alloys was also found to be different from that in the coarse-grained ones [24, 25]. The standard crack growth rate d*a*/d*N* (where *a* is the crack length, *N* is the cycle number) versus stress intensity factor range $\Delta K = K_{max}-K_{min}$ plots for the nanostructured materials showed the same stages of crack propagation as in the coarse-grained materials. However, the fatigue "threshold" ΔK_{th} corresponding to the low-end limit of the curves decreased after SPD processing in Al materials [31].

SPD processing can lead to improvement of the LCF properties in Mg alloys [32]. For example, ECAP processing (for 1–2 passes) of the AZ31B alloy led to significant improvement of the LCF life due to grain refinement as well as reorientation of grains, although it was reduced with further ECAP pressing. The orientation dependence (anisotropy) of the LCF properties in this material was also demonstrated. Similar results were reported for CP Ti [33], where ECAP processing reduced ΔK_{th} and led to the higher cyclic crack growth rate. A lower crack growth rate and a lower fatigue limit due to texture softening and reduction of crack nucleation resistance were also reported for the ECAP-processed Mg alloy.

Analysis of literature on fatigue performance of UFG metallic materials shows that microstructural design is the main approach to improve their fatigue behavior [24, 25]. Microstructure of these materials can be tuned via optimization of the SPD processing parameters, such as temperature, strain, strain rate, and strain path. Intelligent microstructural design can improve HCF properties dramatically. For example, a large enhancement in fatigue life by one order of magnitude in HCF regime was achieved in the AA6061 alloy after a single ECAP pass leading to the formation of subgrains. However, this effect disappeared after the alloy was pressed further for 4 passes due to significant increase of HAGB fraction. On the contrary, the AA2124 showed a higher fatigue life after 8 ECAP passes than that after 4 ECAP passes, and complex microstructure consisting of grains/subgrains was observed in both conditions [34]. Microstructure evolution during SPD processing is determined to a large extent by chemical composition of the material. Therefore, microstructure and fatigue properties can be also tuned via appropriate selection of kind and amount of alloying elements in the alloys.

Enhancement of ductility of nanomaterials is particularly useful for improvement of properties in the LCF regime, where plastic strain amplitudes dominate the fatigue life. As mentioned in Sect. 2.1.2, the ductility of nanostructured metallic materials is often enhanced via their heat treatment after SPD which results in decrease of dislocation density in the interior of nanograins, as well as in relaxation of non-equilibrium grain boundaries. For example, this approach was successfully applied to nanostructured CP Ti [26]. Very promising fatigue properties in the LCF and HCF regimes can be also achieved by introducing the so-called bimodal grain size distributions, where nanograins having the size of several hundred nanometers are embedded into a matrix of coarse grains having the size of several microns [25]. Such bimodal microstructures can be generated via heat treatments of nanostructured metallic materials, although it cannot be done in every material.

Cyclic stability and fatigue life can be also enhanced via manipulation with shear banding. On the one hand, it was suggested that prevention of macroscopic shear banding should improve fatigue life [24]. This can be done, for example, via stabilization of nanostructures against growth of nanograins during cyclic deformation, as coarse grains can act as sites for the formation of shear bands. On the other hand, higher fatigue ductility and consequently higher fatigue life of cryorolled CP Al coincided with a much higher density of shear bands [35]. It was shown that grain coarsening within the macroscopic shear bands increases the capacity of these shear bands to accommodate strain by eliminating the heavily worked regions from the shear bands, allowing additional strain to be accommodated in those regions. These large strain-free grains are softer than their ultrafine-grained surroundings, and this encourages further localization of deformation within the shear bands. However, such effect may not be expected in other nanostructured materials. Special attention should be paid to the as-SPD nanostructures formed in the material, since macroscopic shear bands can grow from shear bands induced into material during SPD processing.

2.2 Properties at High Temperatures

2.2.1 Creep Behavior

Creep is a property which denotes the plastic flow of a material when subjected to a constant stress or a constant load over a long period of time where this stress or load is below the fracture stress. When crystalline materials deform by creep, they exhibit characteristic flow properties in which there is an instantaneous initial strain and then a primary region of creep where the creep rate continuously decreases in stage I, a secondary or steady-state region of creep where the creep rate remains reasonably constant in stage II, and then a tertiary region where the rate of creep accelerates up to final fracture in stage III. Generally, in creep experiments and in many practical applications, the emphasis is placed on the occurrence of flow in stage II, where very significant strains may be attained prior to the onset of stage III.

It is now well-established that the flow behavior in creep may be expressed in a simple relationship that applies, at least in a general form, for all creep mechanisms. This means that it is appropriate to conduct creep experiments to measure the relevant properties and then to use the creep relationship for a direct comparison with the predictions of various theoretical mechanisms. The relationship for the steady-state creep rate, $\dot{\varepsilon}$, may be expressed in a simple form given by [36]

$$\dot{\varepsilon} = \frac{ADGb}{kT}\left(\frac{b}{d}\right)^{p}\left(\frac{\sigma}{G}\right)^{n} \tag{2.3}$$

where D is the appropriate diffusion coefficient which defines the creep flow mechanism where it is equal to D_0 exp ($-Q/RT$) where D_0 is a frequency factor, Q is the activation energy for the diffusive process, R is the gas constant and T is the absolute temperature, G is the shear modulus of the material, **b** is the Burgers vector, k is Boltzmann's constant, d is the grain size, σ is the applied stress, p and n are the exponents of the inverse grain size and the stress, respectively, and A is a dimensionless constant. It follows immediately from inspection of Eq. (2.3) that the relationship contains, for any specific testing conditions, only a limited number of variables. Specifically, the creep behavior may be expressed explicitly by evaluating the values of the four different parameters given by Q, n, p and the constant A. This means, in practice, that laboratory experiments are generally conducted to measure these various quantities for any selected material.

Creep occurs either through the movement of dislocations within the crystalline matrix or by the diffusion of vacancies. Under many testing conditions, microstructural observations after creep testing reveal the division of the grains into arrays of subgrains where the subgrain boundaries are interfaces having low angles of misorientation. It was established in early experiments on many metals that the average subgrain size, λ, in the steady-state region is given by a relationship of the form [37]

$$\frac{\lambda}{\mathbf{b}} = \zeta \left(\frac{\sigma}{G} \right)^{-1} \tag{2.4}$$

where ζ is a constant having a value close to ~20. Later, it was shown that Eq. (2.4) applies equally in ceramic materials [38], and therefore, it may be considered as a fundamental relationship in the high temperature flow properties of all crystalline materials.

In coarse-grained polycrystalline materials, the creep strain is usually produced through the intragranular movement of dislocations where these dislocations arise from the generation of dislocation loops at places such as Frank–Read sources. In practice, these sources are located on different planes and the loops spread outwards in concentric circles so that ultimately there is an interference between the stress fields of dislocations on different but parallel slip planes. Under these conditions, the leading screw dislocations have different signs on the two adjacent planes so that they cross-slip and annihilate, but the edge dislocations are effectively blocked so that they pile up until the back stress is relieved by the two leading dislocations climbing together and annihilating. When this occurs, an additional loop may be generated by the Frank–Read source so that the process is regenerative and the whole process may produce a large strain within the material. Since the effective rate-controlling step is the climb of dislocations, this means that the process is dependent upon the emission and absorption of vacancies and the activation energy for creep is therefore Q_ℓ which corresponds to the appropriate value for lattice self-diffusion. In this creep mechanism, the stress exponent in Eq. (2.3) is given by $n \approx 4.5$ with the precise value dependent upon the stacking-fault energy of the material. Since this process occurs intragranularly, there is no dependence on grain size and $p = 0$ in Eq. (2.3).

This creep mechanism involves the two steps of dislocation glide and then dislocation climb and it is reasonable to assume, at least in most materials, that the rate of climb is much slower than the rate of glide so that the climb process becomes rate-controlling. Nevertheless, in metallic alloys, it is possible that solute atoms may form impurity atmospheres around the dislocations and this can introduce viscous drag where the solute atmospheres are dragged forward in the glide step so that the glide process becomes slower than climb. Under these conditions, the stress exponent is reduced to $n = 3$, there is again no dependence on grain size so that $p = 0$ and the activation energy for this process is now Q_i which corresponds to the appropriate value for interdiffusion of the solute. Experiments show that in many alloys, there is a well-defined transition from control by dislocation climb at lower stresses to control by dislocation glide at higher stresses and a relationship is now available predicting the precise value of this transition for any material and experimental condition [39]. Finally, there is the potential for the dislocations to break away from their solute atmospheres at very high stresses to give $n > 3$ in the region of power-law breakdown [40].

The intragranular movement of dislocations through glide and climb is a major flow process in all crystalline materials, but there is also another important creep process in the form of diffusion creep, where flow occurs through the movement of vacancies acting in response to the applied stress. There are two different and distinct types of diffusion creep and they both depend upon the creation of an excess vacancy concentration along those grain boundaries lying perpendicular to the tensile axis and a corresponding vacancy depletion along those grain boundaries experiencing a compressive stress. This difference in vacancy concentration produces a stress-directed flow of vacancies in an attempt to produce a reasonably equilibrated situation. In Nabarro–Herring diffusion creep [41, 42], the vacancy flow occurs through the crystalline matrix and it leads to $n = 1, p = 2$ and an activation energy of Q_ℓ, whereas in Coble diffusion creep [43], the vacancy flow occurs along the grain boundaries and $n = 1, p = 3$ and the activation energy is Q_{gb} for grain boundary diffusion. It is important to note that diffusion creep is dependent upon the presence of grain boundaries and therefore the flow processes depend on grain size so that both mechanisms become increasingly important as the grain size is reduced. This is, therefore, an important consideration when considering the flow properties of bulk nanostructured materials.

2.2.2 Achieving Superplasticity in Tensile Testing

When polycrystalline materials are pulled in tension, the elongations to failure are generally fairly small. However, early experiments showed that it was possible, under some conditions at least, to achieve quite high tensile elongations prior to failure. The first major achievement in this field was the work of Pearson in the United Kingdom in 1934, when a tensile elongation of ~1950% was reported in a near-eutectic Bi–Sn alloy [44]. It is important to note that the elongation achieved in tension increases directly with an increase in the strain rate sensitivity, m, where m is defined as $(\partial \ln \sigma / \partial \ln \dot{\varepsilon})$ where σ is the flow stress [45]. Detailed experiments show that superplasticity requires a strain rate sensitivity close to $m \approx 0.5$ [46] and this is higher than the values of $m \approx 0.2$ associated with control by dislocation climb and $m \approx 0.3$ for control by dislocation glide or solute drag. Nevertheless, the high strain rate sensitivity of $m \approx 0.3$ in dislocation glide means that this mechanism has a potential for producing reasonably high elongations, of the order of up to ~300%, and this requires that superplasticity is defined formally as an elongation of at least 400% and with a strain rate sensitivity close to $m \approx 0.5$ [47].

The occurrence of superplastic flow in tensile testing is possible only if two specific requirements are fulfilled [46]. First, it is necessary to have a small and stable grain size, which is typically less than ~10 μm. Second, since superplasticity is a diffusion-controlled process, it requires a testing temperature at or above ~0.5 T_m where T_m is the absolute melting temperature of the material. In practice, these two requirements are incompatible because small grains will increase in size at elevated temperatures, but grain growth may be avoided by using two-phase eutectic or eutectoid alloys or by introducing a fine dispersion of a second phase to act as a grain refiner. Generally, superplastic grain sizes are achieved by subjecting materials to appropriate thermo-mechanical treatments, but these procedures are suitable only for obtaining grain sizes that are within the micrometer range. Processing through the application of severe plastic deformation provides an opportunity for achieving much smaller grain sizes in the submicrometer or even the nanometer range and these materials are therefore ideal for achieving excellent superplastic properties.

An example of this effect is shown in Fig. 2.11, where an exceptionally high elongation of 3050% was achieved in an Mg–5.5% Zn–0.5% Zr alloy after preparing by extrusion and then processing by ECAP through 2 passes at a temperature of 473 K with the tensile testing also conducted at 473 K using an initial strain rate of 1.0×10^{-4} s^{-1} [48].

Early experiments in conventional superplasticity, using samples of a two-phase Zn–22% Al eutectoid alloy, showed that for individual tests conducted at a constant temperature, the measured elongations to failure reached maximum values at intermediate strain rates, but the elongations decreased at both slower and faster strain rates [49]. This region of maximum superplasticity was termed the superplastic

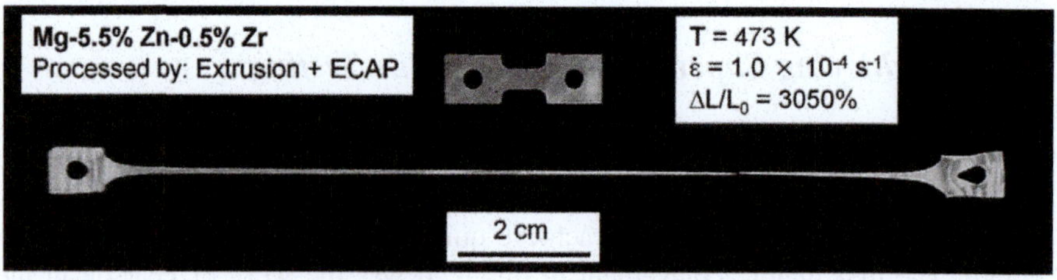

Fig. 2.11 Appearance of a tensile specimen of a magnesium alloy pulled to failure at an elongation of 3050%: the upper specimen is untested. (Reproduced from [48], Copyright 2008, with permission of John Wiley & Sons)

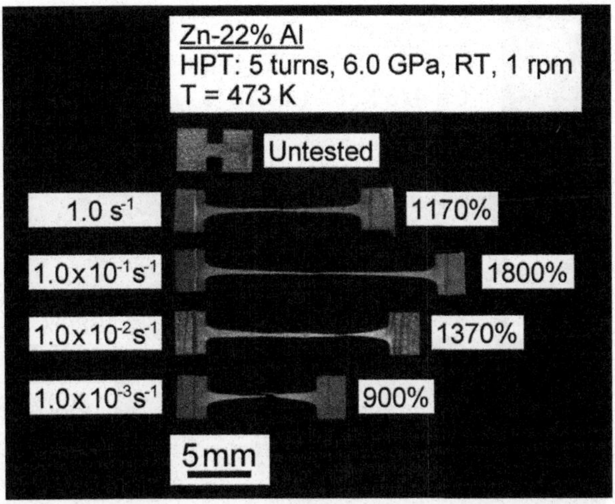

Fig. 2.12 Appearance of specimens of a Zn-22% Al alloy processed by HPT and then pulled to failure at different strain rates at 473 K. (Reprinted from [50], Copyright 2011, with permission from Elsevier)

region II, and the regimes at slower and faster strain rates were designated regions I and III, respectively. This same type of behavior is evident also in samples processed by SPD techniques, as shown in Fig. 2.12 for Zn–22% Al samples processed by HPT at room temperature (RT) under an applied pressure of 6.0 GPa for 5 revolutions at a rotation rate of 1 rpm and then tested in tension at 413 K under initial strain rates from 1.0×10^{-3} s^{-1} to 1.0 s^{-1} [50]. Thus, the maximum elongation of 1800% occurs at a strain rate of 1.0×10^{-1} s^{-1}, which corresponds to the superplastic region II for this material at this testing temperature.

In practice, the elongation achieved in tensile testing depends also on the processing conditions. Figure 2.13 shows an example for the same Zn–22% Al alloy processed by HPT through 1, 3, and 5 revolutions, and these three samples have grain sizes of ~440, ~370, and ~ 350 nm, respectively [50]. The increase in elongation with increasing numbers of torsional revolutions is therefore consistent with the reduction in grain size.

2.2.3 *Mechanism of Superplasticity in Ultrafine-Grained Materials*

An important characteristic of all superplastic flow is that the samples pull out to very high elongations, but nevertheless the individual grains within the polycrystalline matrix remain reasonably equiaxed. This suggests that the deformation occurs through grain boundary sliding, in which the individual grains move over each other under the action of the applied stress [51]. It is now well established that grain boundary sliding may account for essentially all of the strain under optimum

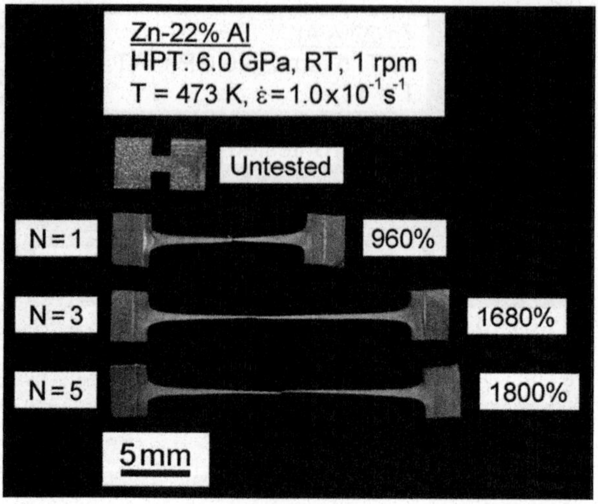

Fig. 2.13 The variation in the elongation to failure with the numbers of torsional revolutions for a Zn-22% Al alloy processed by HPT for 1, 3, and 5 revolutions and then pulled to failure at an initial strain rate of 1.0×10^{-1} s^{-1} at 473 K. (Reprinted from [50], Copyright 2011, with permission from Elsevier)

superplastic conditions [52], but, in order to prevent the opening up of holes or cavities, this sliding must be accommodated by a limited amount of intragranular slip. The occurrence of dislocation slip as an accommodation mechanism has been demonstrated in several experiments including in a Pb–62% Sn eutectic alloy tested under superplastic conditions [53]. This means, therefore, that it is necessary to include this slip process within any development of the deformation mechanism in superplastic flow.

In order to develop a physical mechanism for superplasticity, it is convenient to examine the process of grain boundary sliding both in coarse-grained materials and in superplastic materials having very small grain sizes. The fundamental principles for these two conditions are illustrated schematically in Fig. 2.14 [19]. For coarse-grained materials, as in Fig. 2.14a, sliding on a grain boundary leads to a pile-up of dislocations at triple points as at A, and the consequent stress concentration is accommodated by intragranular slip in the next grain so that the dislocations move across the grain and pile-up at the first subgrain boundary as at B. For superplastic materials as in Fig. 2.14b, the grains are sufficiently small that, as demonstrated in a very early analysis [55], no subgrains are formed during tensile flow. This means that the stress concentration from grain boundary sliding due to the pile-up of dislocations at the triple point C are now accommodated by intragranular dislocations which move across the next grain and pile up at the opposite grain boundary at D. Both of these processes lead to a flow relationship of the form shown in Eq. (2.3), but for the coarse-grained materials in Fig. 2.14a, the relevant terms are $p = 1$, $n = 3$, $Q = Q_\ell$ and $A \approx 10^3$, whereas for superplastic materials in Fig. 2.14b, the terms are $p = 2$, $n = 2$, $Q = Q_{gb}$ and $A \approx 10$ where the difference arises because of the change in the flow length for the accommodating slip. Thus, in non-superplastic flow, the

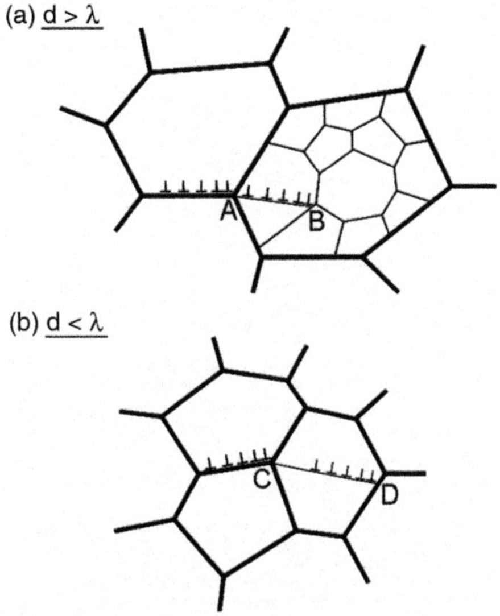

Fig. 2.14 The principles of grain boundary sliding in (**a**) a coarse-grained material where subgrains form and (**b**) a superplastic material where the grains are too small for the formation of any subgrains. (Reprinted from [54], Copyright 2011, with permission from Elsevier)

accommodation occurs over a distance which is equal to the subgrain size as given in Eq. (2.4), and it is a simple process to incorporate this length into an analysis of the flow properties.

Recent analyses show that the predictions of this mechanism for superplastic flow are in very good agreement with the experimental data reported for superplastic metals processed by ECAP and HPT. Taking Al and Mg alloys as examples, where numerous superplastic results are now available, the experimental data for these two materials are plotted in the form of the temperature and grain size compensated strain rate against the normalized stress in Fig. 2.15a, b, respectively [56]. In both plots, the solid line marked $\dot{\varepsilon}_{sp}$ denotes the theoretical prediction from this model and this line has a slope of $n = 2$. Separate sets of datum points are shown for materials processed by ECAP in black and HPT in red, and then these two sets are further designated and separately identified by drawing encircling ovals in blue and pink for processing by ECAP and HPT, respectively. It is readily apparent that there is excellent agreement between the experimental datum points and the theoretical line for superplastic flow, and this confirms that the flow mechanism in these ultrafine-grained materials prepared by SPD processing agrees well with the predictions for conventional superplasticity. It is also apparent that the points for the HPT samples in both materials are shifted slightly toward higher stresses compared with the theoretical prediction. This small shift arises because of the experimental difficulties of using the extremely small tensile samples produced by HPT processing in order to measure the flow properties. Nevertheless, the overall agreement between experiment and theory is generally excellent.

Deformation mechanism maps represent a convenient procedure for pictorially documenting creep mechanisms and for using as a predictive tool in order to make reasonable estimates of the rate-controlling creep processes and the inherent strain

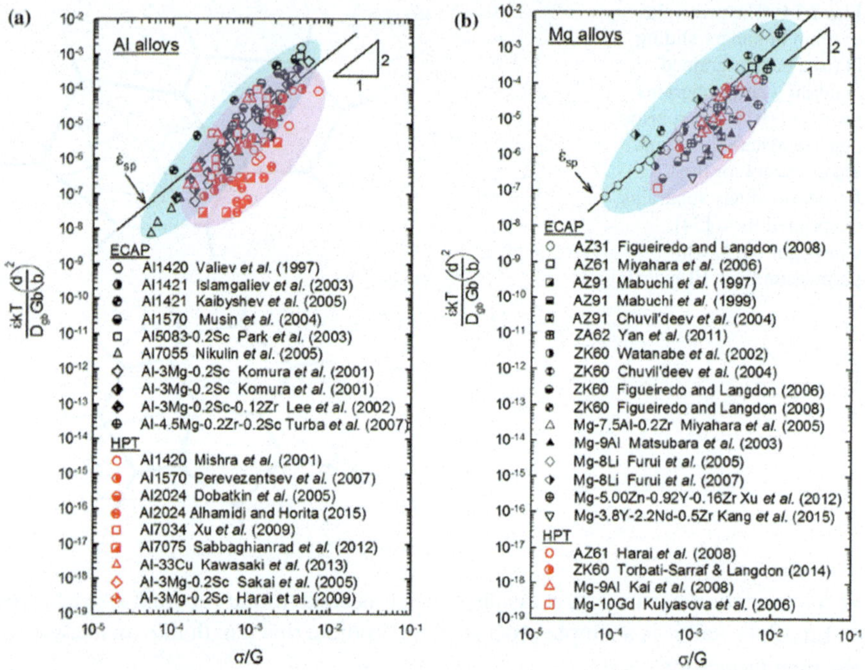

Fig. 2.15 Temperature and grain size compensated strain rate versus normalized stress for ultrafine-grained samples processed by ECAP in black and HPT in red for (**a**) Al alloys and (**b**) Mg alloys: the solid line designates the theoretical prediction for superplastic flow and the encircling ovals are in blue for ECAP and pink for HPT. (Reprinted by permission from Springer Nature [56], Copyright 2016)

rates associated with these mechanisms under different testing conditions [57]. An example of a deformation mechanism map is shown in Fig. 2.16, where the two axes represent the normalized grain size, $d/\mathbf{b}$, and the normalized stress, σ/G, plotted at an absolute temperature of 473 K and showing the three regions associated with superplastic materials, labelled regions I, II, and III, and the theoretical predictions for the occurrence of Nabarro–Herring and Coble creep [57]. The dashed line corresponds to the prediction of Eq. (2.4) where the grain size is set equal to the subgrain size. This dashed line corresponds reasonably to the transition from region III at high strain rates to the superplastic region II at intermediate strain rates, and this confirms that superplastic flow requires a grain size that is too small for the formation of any subgrains within the grains. The four experimental points for different strain rates shown in Fig. 2.16 correspond to the samples processed by HPT through 5 turns at 473 K using an applied pressure of 6.0 GPa, as shown earlier in Fig. 2.12. Representations of this kind provide a simple procedure for making predictions on the dominant rate-controlling flow process under different experimental conditions [57].

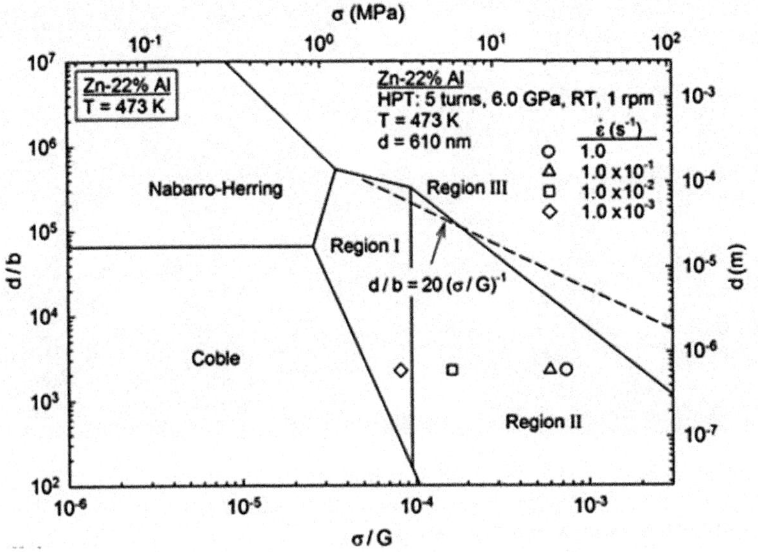

Fig. 2.16 A deformation mechanism map for the Zn-22% Al alloy using experimental data for the four samples processed by HPT and tested in tension at different strain rates in Fig. 2.14. (Reprinted by permission from Springer Nature [57], Copyright 2013)

2.2.4 Developing a Superplastic-Forming Capability

The occurrence of superplastic flow in laboratory experiments was considered initially as little more than a scientific curiosity, but later it was demonstrated that the advent of superplasticity provides an opportunity for forming complex parts, often in the form of beautifully curved panels, that are useful in a very wide range of industrial applications [58]. This gave rise to the superplastic forming industry that is now used widely for the fabrication of parts for many applications ranging from the bodies of high-performance sports cars to curved parts for aerospace applications, decorative architectural panels, and boxes and containers for computers and other accessories. Furthermore, the introduction of SPD processing, and the consequent production of metals with submicrometer and nanometer grain sizes, have enhanced the opportunities for making use of superplastic forming and provided new procedures that are now available from using these materials having exceptional grain refinement [59].

By comparison with conventional superplastic metals, materials processed by SPD techniques have smaller grain sizes and this means that, since the strain rate in superplastic flow varies inversely with the grain size raised to a power of 2, the introduction of exceptional grain refinement leads to the occurrence of superplasticity at faster strain rates. In conventional superplastic forming, a plate of a superplastic alloy is placed on a heated platen within a furnace and then held in place at an elevated temperature for a period of typically ~20–30 min so that it is formed into the desired shape at a strain rate of the order of ~10^{-3} to 10^{-2} s^{-1}. This means that the

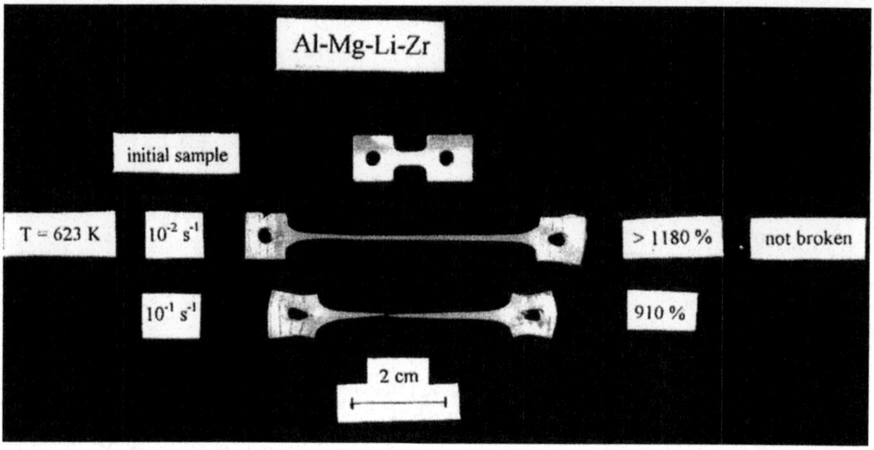

Fig. 2.17 Samples of an Al-Mg-Li-Zr alloy showing the opportunity for achieving high strain rate superplasticity after producing an ultrafine grain size by ECAP processing. (Reprinted from [60], Copyright 1997, with permission from Elsevier)

process is inherently slow, and therefore, it is not possible to form large numbers of parts in a single session. It follows therefore that if the strain rate in superplasticity is increased to ~10^{-1} to 1 s^{-1} the time involved in the forming process will be reduced to ~20–30 s, and this will make the process effective for the processing of parts using fairly rapid mass production.

A first demonstration of the rapid strain rates involved in the superplastic tensile testing of SPD materials occurred in 1997 using two commercial aluminum alloys processed by ECAP where the ultrafine grain sizes were effectively retained through the presence of precipitates. An example is shown in Fig. 2.17 for an Al–Mg–Li–Zr alloy, where the top sample is untested, the lowest sample was pulled to failure at 910% at 623 K with a strain rate of 10^{-1} s^{-1} and the central specimen pulled out to an elongation of 1180% without failure at a strain rate of 10^{-2} s^{-1} [60]. The appearance of these two tensile specimens demonstrates conclusively that it is possible to achieve high strain rate superplasticity in metals subjected to SPD processing.

Although SPD processing as in ECAP provides materials having the capability for use in rapid forming operations, it is important to recognize that industrial superplastic forming practice is based on the processing of thin metal sheets into complex shapes. In practice, the samples produced by ECAP are generally in the form of rods or bars having either square or round cross-sections. It is necessary to check, therefore, whether these bars will retain their superplastic properties when they are cold rolled into a sheet form after the SPD processing. This was first investigated using a superplastic Al–3% Mg–0.2% Sc alloy where this alloy was selected because it exhibits excellent superplastic properties after ECAP [61]. The result is shown in Fig. 2.18, where it is apparent that both sets of samples exhibit very similar elongations to failure, and therefore, the superplastic properties are not in any way diminished by cold rolling after the SPD operation. This further supports the use of SPD processing in the preparation of materials for superplastic forming operations.

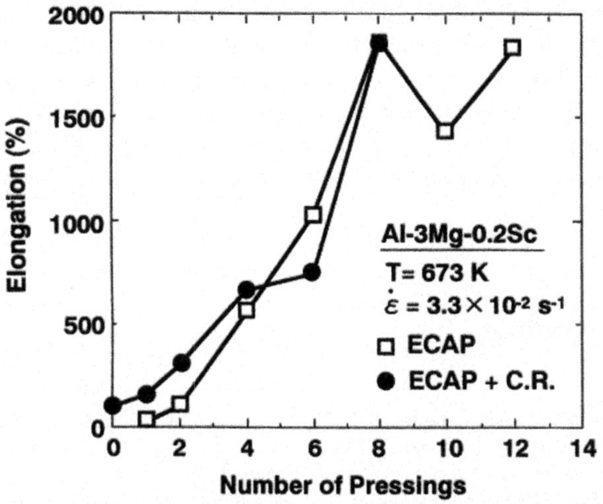

Fig. 2.18 Elongation to failure plotted against the number of passes in ECAP for samples of an Al-Mg-Sc alloy tested at 673 K at a strain rate of 3.3×10^{-2} s^{-1} after processing by ECAP (open points) or processing by ECAP and then cold rolling (CR) (solid points). (Reprinted from [61], Copyright 2001, with permission from Elsevier)

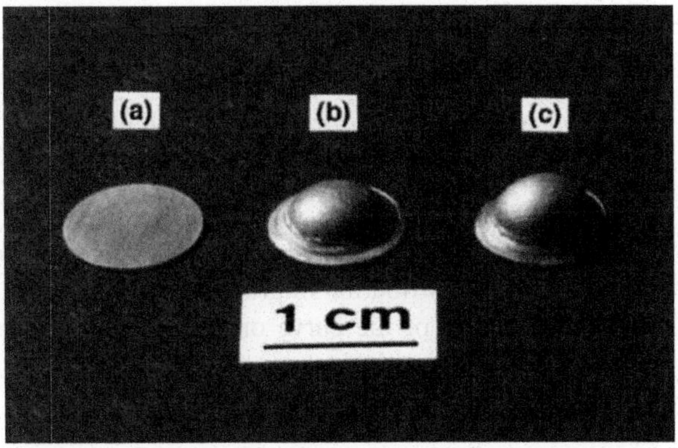

Fig. 2.19 Domes produced by blow forming of an Al-3% Mg-0.2% Sc alloy cut from billets processed by ECAP: (**a**) initial disk without blow forming and disks held at 673 K under a gas pressure of 1.0 MPa for (**b**) 30 and (**c**) 60 s. (Reprinted from [62], Copyright 2001, with permission from Elsevier)

A first direct illustration of high-speed superplastic forming in a sample prepared by an SPD technique was presented in 2000 and the results are illustrated in Fig. 2.19 [62]. In this research, an Al–3% Mg–0.2% Sc alloy in the form of a rod was initially processed by ECAP and then disks were cut with thicknesses of ~0.3 mm perpendicular to the longitudinal axis of the billet. An example of the cut disk is shown in Fig. 2.19 at (a). The disks were then placed in a biaxial gas-pressure superplastic forming facility wherein the disks were clamped around the peripheries and then subjected on the lower surfaces to a constant argon gas pressure for very short periods of time. This gas pressure was set at 1.0 MPa, which is similar to the pressure used in conventional superplastic forming operations. Thus, by employing the same pressure as in regular forming procedures it means that the process can be conducted easily by the superplastic forming industry without the need for any expensive re-tooling. The excellent domes that were achieved using this procedure are shown in Fig. 2.19 after holding in the facility for times of (b) 30 and (c) 60 s, respectively. This result provides an example, therefore, of the excellent rapid superplastic forming that may be achieved in ultrafine-grained materials after SPD processing.

Suggested Reading

1. Valiev RZ, Zhilyaev AP, Langdon TG (2014) Bulk nanostructured materials: fundamentals and applications. Wiley, Hoboken, p 456
2. Kolobov YR, Valiev RZ et al (2007) Grain boundary diffusion and properties of nanostructured materials. CISP: Cambridge International Science Publishing, Cambridge, p 250
3. Kassner ME (2015) Fundamentals of creep in metals and alloys. Elsevier, Amsterdam, p 356

Problems and Tasks

2.1 Mechanical Properties at Ambient Temperature

2.1.1 *Superstrength and Hardening Mechanisms*

1. Describe and analyze the peculiarities of the Hall–Petch relation in the case of UFG materials.
2. What are the reasons for the formation of superstrong states in UFG materials?
3. What is the contribution of grain boundary segregations to the formation of the negative slope of the curve on the diagram corresponding to the Hall–Petch relation?

2.1.2 *Ductility and Strategies for Its Improvement*

4. What are the differences between the terms "uniform elongation" and "elongation to failure" and what processes control their meanings?
5. What are the trends and reasons for decreasing ductility, while increasing the strength of materials subjected to conventional plastic deformation techniques?
6. Analyze "mechanical" strategies for increasing ductility of metallic materials.
7. Analyze "microstructural" strategies for increasing ductility of metallic materials.
8. Write down and explain the meaning of the equation corresponding to the Considère criterion and define all quantities entering into it. For predicting the homogeneous deformation of which materials is this criterion suitable?
9. Write down and explain the meaning of the equation corresponding to the Hart criterion and define the quantities entering into it. For predicting the homogeneous deformation of which materials is this criterion suitable?

10. What is the essence and what are the reasons for the paradox of plastic deformation of UFG materials obtained by SPD?
11. Analyze the role of the value of the coefficient of velocity sensitivity of the flow stress in the level of plasticity of metallic materials.
12. What is the relationship between the proportion of high angle grain boundaries, specific deformation mechanisms, and plasticity of UFG metals and alloys?
13. Analyze the role of bimodal grain structure components in increasing the ductility of UFG metallic materials.
14. What is the contribution and what are the reasons for the influence of nanoprecipitations, nanoclusters, nanotwins on the values of strength and ductility of UFG materials?

2.1.3 *Fatigue Behavior*

15. What are the causes of the differences between the multicycle fatigue behavior of UFG materials and that of their coarse-grained counterparts?
16. Analyze the reasons affecting the low-cycle fatigue behavior of UFG materials.
17. Suggest promising approaches to improve the fatigue behavior of UFG materials.

2.2 Properties at High Temperatures

2.2.1 *Creep Behavior*

18. Describe mechanisms of dislocation creep implementation.
19. Describe types of diffusive creep and the role of grain size in creep implementation.

2.2.2 *Achieving Superplasticity in Tensile Testing*

20. Analyze the conditions of the superplasticity effect in metallic materials.
21. Describe and explain the effect of reduced grain size on the occurrence of the superplasticity effect in metallic materials.

2.2.3 *Mechanism of Superplasticity in Ultrafine-Grained Materials*

22. Describe the features of grain boundary sliding in coarse grained and UFG materials.

References

1. Valiev RZ, Islamgaliev RK, Alexandrov IV (2000) Bulk nanostructured materials from severe plastic deformation. Prog Mater Sci 45:103–189
2. Morris DG (2011) Strengthening mechanisms in nanocrystalline metals. In: Whang SH (ed) Nanostructured metals and alloys. Processing, microstructure, mechanical properties and applications. Woodhead Publishing Limited, Cambridge, pp 297–325
3. Pande C, Cooper K (2009) Nanomechanics of Hall–Petch relationship in nanocrystalline materials. Prog Mater Sci 54:689–706

4. Valiev RZ, Enikeev NA, Langdon TG (2011) Towards superstrength of nanostructured metals and alloys produced by SPD. Kovove Mater 49:1–9
5. Firstov SA, Rogul TG, Shut OA (2009) Transition from microstructures to nanostructures and ultimate hardening. Funct Mater 16:4
6. Valiev RZ, Enikeev NA, Murashkin MY, Kazykhanov VU, Sauvage X (2010) On the origin of extremely high strength of ultrafine-grained Al alloys produced by severe plastic deformation. Scr Mater 63:949–952
7. Li JCM (2011) Grain Boundaries in Nanocrystalline Materials. In: Li JCM (ed) Mechanical properties of nanocrystalline materials. Pan Stanford Publishing, Singapore, p 193
8. Liddicoat PV, Liao XZ, Zhao YH, Zhu YT, Murashkin MY, Lavernia EJ, Valiev RZ, Ringer SP (2010) Nanostructural hierarchy increases the strength of aluminium alloys. Nat Commun 1:1–7
9. Kim JG, Enikeev NA, Seol JB, Abramova MM, Karavaeva MV, Valiev RZ, Park CG, Kim HS (2018) Superior strength and multiple strengthening mechanisms in nanocrystalline TWIP steel. Sci Rep 8:1–10
10. Raabe D, Choi PP, Li Y, Kostka A, Sauvage X, Lecouturier F, Hono K, Kirchheim R, Pippan R, Embury D (2010) Metallic composites processed via extreme deformation: Toward the limits of strength in bulk materials. MRS Bull 35:982–991
11. Ovid'ko IA, Valiev RZ, Zhu YT (2018) Review on superior strength and enhanced ductility of metallic nanomaterials. Prog Mater Sci 94:462–540
12. Zhao YH, Zhu YT, Lavernia EJ (2010) Strategies for improving tensile ductility of bulk nanostructured materials. Adv Eng Mater 12:769–778
13. Ma E (2006) Eight routes to improve the tensile ductility of bulk nanostructured metals and alloys. JOM 58:49–53
14. Valiev RZ, Alexandrov IV, Lowe TC, Zhu YT (2002) Paradox of strength and ductility in metals processed by severe plastic deformation. J Mater Res 17:5–8
15. Alawadhi MY, Sabbaghianrad S, Huang Y, Langdon TG (2021) Evaluating the paradox of strength and ductility in ultrafine-grained oxygen-free copper processed by ECAP at room temperature. Mater Sci Eng A 802:140546
16. Valiev RZ, Murashkin MY, Kilmametov AR, Straumal BB, Chinh NQ, Langdon TG (2010) Unusual super-ductility at room temperature in an ultrafine-grained aluminum alloy. J Mater Sci 45:4718–4724
17. Wang Y, Chen M, Zhou F, Ma E (2002) High tensile ductility in a nanostructured metal. Nature 419:912–915
18. Zhao YH, Liao XZ, Cheng S, Ma E, Zhu YT (2006) Simultaneously increasing the ductility and strength of nanostructured alloys. Adv Mater 18:2280–2283
19. Horita Z, Ohashi K, Fujita T, Kaneko K, Langdon TG (2005) Achieving high strength and high ductility in precipitation-hardened alloys. Adv Mater 17:1599–1602
20. Zhao YH, Bingert JE, Liao XZ, Cui BZ, Han K, Sergueeva AV, Mukherjee AK, Valiev RZ, Langdon TG, Zhu YT (2006) Simultaneously increasing the ductility and strength of ultrafine-grained pure copper. Adv Mater 18:2949–2953
21. Wang Y, Ma E, Valiev RZ, Zhu Y (2004) Tough nanostructured metals at cryogenic temperatures. Adv Mater 16:328–331
22. Zhu YT, Liao XZ, Wu XL (2012) Deformation twinning in nanocrystalline materials. Prog Mater Sci 57:1–62
23. Suresh S (1998) Fatigue of materials, 2nd edn. Cambridge University Press, Cambridge, p 679
24. Hoeppel HW, Goeken M (2011) Fatigue behavior in nanostructured metals. In: Whang SH (ed) Nanostructured metals and alloys: processing, microstructure, mechanical properties and applications. Woodhead Publishing Limited, p 507
25. Estrin Y, Vinogradov A (2010) Fatigue behavior of light alloys with ultra-fine grain structure produced by severe plastic deformation: an overview. Int J Fatigue 32:898–907
26. Semenova IP, Salimgareeva GK, Latysh VV, Lowe TC, Valiev RZ (2009) Enhanced fatigue strength of commercially pure Ti processed by severe plastic deformation. Mater Sci Eng A 503:92–95

27. Saitova LR, Höppel HW, Göken M, Semenova IP, Valiev RZ (2009) Cyclic deformation behavior and fatigue lives of ultrafine-grained Ti-6AL-4V ELI alloy for medical use. Int J Fatigue 31:322–331
28. Hoeppel HW, May J, Prell M, Goeken M (2011) Influence of grain size and precipitation state on the fatigue lives and deformation mechanisms of CP aluminium and AA6082 in the VHCF-regime. Int J Fatigue 33:10–18
29. Malekjani S, Hodgson PD, Cizek P, Hilditch TB (2011) Cyclic deformation response of ultra-fine pure Al. Acta Mater 59:5358–5367
30. Mughrabi H, Hoeppel HW (2001) Cyclic deformation and fatigue properties of ultrafine grain size materials: current status and some criteria for improvement of the fatigue resistance. Mater Res Soc Symp Proc B2.1.1:634
31. Huebner P, Kiessling R, Biermann H, Hinkel T, Jungnickel W, Kawalla R, Hoeppel HW, May J (2007) Static and cyclic crack growth behavior of ultrafine-grained Al produced by different severe plastic deformation methods. Metall Mater Trans A 38:1926–1933
32. Wu L, Stoica GM, Liao HH, Agnew SR, Payzant EA, Wang G, Fielden DE, Chen L, Liaw PK (2007) Fatigue-property enhancement of magnesium alloy AZ31B through ECAP. Metall Mater Trans A 38:2283–2289
33. Hanlon T, Tabachnikova ED, Suresh S (2005) Fatigue behavior of nanocrystalline metals and alloys. Int J Fatigue 27:1147–1158
34. Lapovok R, Loader C, Dalla Torre FH, Semiatin SL (2006) Microstructure evolution and fatigue behavior of 2124 aluminum processed by ECAE with back pressure. Mater Sci Eng A 425:36–46
35. Malekjani S, Hodgson PD, Stanford NE, Hilditch TB (2013) The role of shear banding on the fatigue ductility of ultrafine-grained aluminium. Scr Mater 68:269–272
36. Langdon TG (2020) A lifetime of research in creep, superplasticity and ultrafine-grained materials. Adv Eng Mater 22:1900442
37. Bird JE, Mukherjee AK, Dorn JE (1969) Correlations between high-temperature creep behavior and structure. In: Brandon DG, Rosen A (eds) Quantitative relation between properties and microstructure. Israel Universities Press, Jerusalem, pp 255–341
38. Cannon WR, Langdon TG (1988) Review: Creep of ceramics, Part 2 an examination of flow mechanisms. J Mater Sci 23:1–20
39. Mohamed FA, Langdon TG (1974) The transition from dislocation climb to viscous glide in creep of solid solution alloys. Acta Metall 22:779–788
40. Yavari P, Langdon TG (1982) An examination of the breakdown in creep by viscous glide in solid solution alloys at high stress levels. Acta Metall 30:2181–2196
41. Nabarro FRN (1948) Deformation of crystals by the motion of single ions. In: Report of a conference on strength of solids. The Physical Society, London, pp 75–90
42. Herring C (1950) Diffusional viscosity of a polycrystalline solid. J Appl Phys 21:437–445
43. Coble RL (1963) A model for boundary diffusion controlled creep in polycrystalline materials. J Appl Phys 34:1679–1682
44. Pearson CE (1934) The viscous properties of extruded eutectic alloys of lead-tin and bismuth-tin. J Inst Metals 54:111–124
45. Woodford DA (1969) Strain-rate sensitivity as a measure of ductility. Trans ASM 62:291–293
46. Langdon TG (1982) The mechanical properties of superplastic materials. Metall Trans A 13A:689–701
47. Langdon TG (2009) Seventy-five years of superplasticity: historic developments and new opportunities. J Mater Sci 44:5998–6010
48. Figueiredo RB, Langdon TG (2008) Record superplastic ductility in a magnesium alloy processed by equal-channel angular pressing. Adv Eng Mater 10:37–40
49. Ishikawa H, Mohamed FA, Langdon TG (1975) The influence of strain rate on ductility in the suoperplastic Zn-22% Al eutectoid. Philos Mag 32:1269–1271
50. Kawasaki M, Langdon TG (2011) Developing superplasticity and a deformation mechanism map for the Zn-Al eutectoid alloy processed by high-pressure torsion. Mater Sci Eng A 528:6140–6145

51. Langdon TG (2006) Grain boundary sliding revisited: Developments in sliding over four decades. J Mater Sci 41:597–609
52. Langdon TG (1994) An evaluation of the strain contributed by grain boundary sliding in superplasticity. Mater Sci Eng A 174:225–230
53. Valiev RZ, Langdon TG (1993) An evaluation of the role of intragranular dislocation strain in the superplastic Pb-62% Sn eutectic alloy. Acta Metall Mater 41:949–954
54. Langdon TG (1994) A unified approach to grain boundary sliding in creep and superplasticity. Acta Metall Mater 42:2437–2443
55. Mohamed FA, Langdon TG (1976) Deformation mechanism maps for superplastic materials. Scr Metall 10:759–762
56. Kawasaki M, Langdon TG (2016) Achieving superplastic properties in ultrafine-grained materials at high temperatures. J Mater Sci 51:19–32
57. Kawasaki M, Langdon TG (2013) The many facets of deformation mechanism mapping and the application to nanostructured materials. J Mater Res 28:1827–1834
58. Barnes AJ (2007) Superplastic forming 40 years and still growing. J Mater Eng Perform 16:440–454
59. Langdon TG (2020) The background to superplastic forming and opportunities arising from new developments. Solid State Phenom 306:1–8
60. Valiev RZ, Salimonenko DA, Tsenev NK, Berbon PB, Langdon TG (1997) Observations of high strain rate superplasticity in commercial aluminum alloys with ultrafine grain sizes. Scr Mater 37:1945–1950
61. Akamatsu H, Fujinami T, Horita Z, Langdon TG (2001) Influence of rolling on the superplastic behavior of an Al-Mg-Sc alloy after ECAP. Scr Mater 44:759–764
62. Horita Z, Furukawa M, Nemoto M, Barnes AJ, Langdon TG (2000) Superplastic forming at high strain rates after severe plastic deformation. Acta Mater 48:3633–3640

Chapter 3
Functional and Multifunctional Properties of UFG Metallic Materials

Nanostructuring of metals by SPD processing provides an opportunity to enhance not only their mechanical properties but also various functional ones associated with new physical and chemical properties. Such multifunctional behavior of UFG materials is quite attractive for their innovation application in engineering and medicine and recently has been the subject of robust research. This chapter considers the features and nature of advanced functional and multifunctional properties observed in these materials.

As was noted in Chap. 1, UFG materials produced by SPD are characterized by not only ultrafine grains but a number of other nanostructural features as well, including non-equilibrium grain boundaries, nanophase particles, grain boundary segregations and precipitates, and others that also markedly affect the material behavior, thus creating the opportunity to achieve novel properties in UFG materials.

After the first works on nanostructuring of metals and alloys by SPD techniques (see [1]), there were initiated active studies of these functional properties and the progress of such works has been considered in a number of recent reviews [1–9] which constitute the basis for this chapter. The reviews available so far usually cover 2–3 novel properties of a nanomaterial. In this section, we consider in detail first the physical, then the chemical and multifunctional properties of UFG materials with particular attention to the relation with the structural features of these materials.

3.1 Physical Properties

The physical properties of UFG materials are closely related to the features of their microstructure. The small grain size in metals and alloys processed by SPD techniques, comparable with the conduction electrons' mean-free path and the size of magnetic domains, as well as the nonequilibrium state of the grain boundaries with introduced extrinsic dislocations, high concentration of deformation vacancies,

R. Z. Valiev et al., *Ultrafine-Grained Materials*, The Minerals, Metals & Materials Series, https://doi.org/10.1007/978-3-031-31729-3_3

increased dislocation density and other factors provide a unique level of physical properties in bulk nanomaterials.

According to early experimental studies carried out by differential calorimetry, UFG materials obtained by SPD techniques are characterized by the excess volume of dislocations associated with the formation of a net of high-angle grain boundaries containing extrinsic dislocations (EDs) [1]. These dislocations produce fields of long-range elastic stresses that result in the displacement of atoms from their lattice sites' positions in the grain boundary regions, thereby enabling the increase of the specific volume per atom in the material. Annealing at elevated temperatures leads to a gradual recovery in the microstructure and reduction of the free volume of grain boundaries due to the transition of nonequilibrium grain boundaries to equilibrium ones. Thus, estimates show the change in the relative volume of the grain boundaries is about 3% in UFG materials with a grain size of about 100 nm during subsequent annealing. This change in volume considerably exceeds the difference in the free volume of the amorphous and crystalline states, which is approximately equal to 1%.

The peculiarities of the UFG structure were also investigated by Möessbauer spectroscopy, testifying to two different subspectra in the ferromagnet spectrum [1]. The parameters of the first, more intensive subspectrum pretty much coincide with the parameters of the spectrum of a coarse-grained ferromagnet. The presence of the second subspectrum is obviously associated with the special state of the atoms at and near grain boundaries. The fraction of such atoms can reach 10%. At the same time, the thickness of the boundary layer (the physical width of the grain boundaries) is estimated to be approximately equal to 10 nm.

The conditions for the formation of high segregations of impurity atoms or atoms of alloying elements in the grain boundary regions are created in UFG metals obtained by SPD (see Chap. 1). The formation of such segregations is evidenced by the results of direct investigations conducted by atomic probe tomography. Consequently, SPD-processed metallic alloys are characterized by an increased ultimate dissolution of the alloying atoms in the matrix. For example, in high-carbon steel, an increase in the concentration of C atoms has been observed at grain boundaries of Fe by about 20 times [10].

Thus, the nanostructuring by SPD processing can lead to a dramatic increase in the relative volume fraction of the grain boundaries and the number of crystal lattice defects as well as introduce the change of the lattice parameter, the growth of static and dynamic atomic displacements and free volume, the increase in the limit dissolution of alloying elements in solid solution, and the formation of high segregations of alloying elements at the grain boundaries. Such features of the UFG materials directly affect the level of their properties.

Some basic physical properties of UFG metals in relation to the peculiarities of their specific microstructure are discussed below.

3.1.1 Electrical Conductivity

The electrical conductivity of materials is significant for the development of electronics, power engineering, and computer technology. In this regard, electrical conductivity is of great importance in the creation of modern conductors having also enhanced strength properties.

It is well known that the high electrical conductivity of metallic materials is determined by the mobility of the electrons weakly bound to the atoms, i.e., electrons on the outer shells of atoms. These electrons, easily detachable from the atoms, are termed as free electrons and form a so-called electron gas in the metal. The thermal agitation of free electrons in the absence of an applied potential difference leads to zero electric current flowing. Under the influence of a potential difference, the conduction electrons are easily involved in an ordered movement and create an electric current.

Electric current in metals experiences resistance from the crystal lattice due to the conduction electron scattering on the lattice ions performing thermal vibrations. The efficiency of conduction electron scattering depends on the type of crystal lattice, chemical, and phase composition of the material under study. Crystallographic defects contribute to electrical resistivity. The specific electrical resistivity for alloys consisting of two phases can be calculated using a weighted summation that takes into account the resistivity of each phase ρ_α and ρ_β as well as their volume fractions V_α и V_β,

$$\rho = \rho_\alpha V_\alpha + \rho_\beta V_\beta. \tag{3.1}$$

With no changes in chemical and phase composition, the resistivity ρ for a particular metallic material, according to Mattiessen's rule, can be represented as a sum of two independent components: the temperature-dependent thermal component ρ_T and the temperature-independent structural component ρ_R:

$$\rho = \rho_T + \rho_R. \tag{3.2}$$

It is known that the electrical resistivity ρ at temperature t (°C), not too close to 0 K, is satisfactorily described by the equation

$$\rho_T = \rho_0 (1 + \alpha t), \tag{3.3}$$

where ρ_0 is the specific electrical resistivity at temperature t = 0 °C and α is the linear coefficient of thermal expansion of the given material.

Conduction electrons scattering at thermal vibrations of atoms and equilibrium thermal vacancies of the crystal lattice at low temperatures make a small contribution to the specific electrical resistivity. However, at intermediate and high temperatures, this contribution becomes predominant as compared to that of temperature-independent component.

An estimate of various microstructure parameters' input to the value of the structural component of the resistivity can be made within the additive approximation using the expression

$$\rho_R = \rho_i + \rho_p + \rho_d + \rho_v + \rho_{gb}, \tag{3.4}$$

where ρ_i – the solid-solution component associated with scattering of conduction electrons on the atoms of alloying elements; ρ_p – the component due to the presence of particles of secondary phases, including Guinier-Preston zones, incoherent inclusions, and coherent precipitations; ρ_d – dislocation component; ρ_v – vacancy component due to the presence of deformation vacancies; ρ_{gb} – grain boundary component of specific electrical resistivity.

The formation of a solid solution is accompanied by appearance of elastic microdistortions in a crystal lattice of the basic element as a result of introduction of atoms of alloying elements (impurity) between atoms of the basic element (matrix), or replacement of atoms of the basic element in the sites of its crystal lattice. This, as well as differences in effective radii of scattering of conduction electrons by atoms of matrix and alloying elements, leads to appearance of component ρ_i of specific resistivity. The value of this component is directly proportional to the relative atomic fraction of the alloying elements.

The contribution of ρ_p to the specific electrical resistivity associated with the scattering of conduction electrons on particles of secondary phases is directly proportional to the square of the radius of the latter and inversely proportional to the volume they take. Thermo-mechanical treatment can result in precipitation of particles from solid solution and their dissolution in it. These processes are accompanied by a change in the morphology as well as in the coherence of the relation between the precipitates and the solid solution.

The contributions of the solid-solution component ρ_i and the component ρ_p associated with scattering at the particles of secondary phases are predominant in the thermally independent component of the resistivity of alloys.

In the case of a two-component alloy, the value of the resistivity component ρ_i depends on the concentration c_i of the alloying element atoms in the solid solution according to the equation

$$\rho_i = A_i c_i \left(1 - c_i\right), \tag{3.5}$$

where A_i – a constant independent of the alloy composition, but determined by the atomic numbers of its constituents.

Plastic deformation increases the number of crystal structure defects that are the centers of conduction electrons scattering. Such defects include deformation vacancies, dislocations, deformation twins, low-angle and high-angle grain boundaries, and pores.

However, the influence of plastic deformation on the value of specific electrical resistivity is significantly lower than that of temperature or alloying. At the same time, SPD results in the formation of bulk nanostructured states characterized by the increased values of the temperature-independent static and temperature-dependent dynamic components of the atomic displacements. As already demonstrated in Chap. 1, SPD is usually accompanied by a significant change in the crystal lattice parameters, an increase in dislocation density as well as deformation vacancy concentration, a decrease in grain size and enhancement in the misorientation angle between neighboring grains, an increase in the ultimate solubility of components, etc. The structural component of specific resistivity is, to a certain extent, affected by these microstructure parameters depending on the type of material and SPD processing regimes.

Cu and its alloys as well as Al and its alloys are considered to be the main electrical materials. The development and use of these materials as modern conductors are traditionally determined by the ratio of their electrical conductivity and strength. The data on specific electrical conductivity and ultimate tensile strength for these materials are summarized in Fig. 3.1a, b [11]. It demonstrates that these parameters significantly differ in the bulk coarse-grained and nanostructured states.

The values of specific electrical resistivity for pure nanostructured Cu, obtained by SPD techniques compared to the coarse-grained state, are higher due to the increased concentration of excess vacancies and dislocations as well as the developed net of grain boundaries. Approaches that form a large number of deformation twins with nanoscale thicknesses in the microstructure are more effective in terms of maintaining high electrical conductivity while increasing the strength. Deformation twins, due to their more symmetrical structure, are much weaker in scattering conduction electrons than grain boundaries.

The formation of the microstructure with high density of nanotwins is realized by the temperature reduction and/or increase of the SPD processing rate. As a result, a combination of high strength (about 700 MPa) and electrical conductivity (up to 80% IACS) has been achieved in some works.

Alloys of the Cu–Cr and Cu–Cr–Zr systems are the most familiar dispersion-hardening Cu alloys. Improved strength and thermal stability of these alloys make them more advantageous than pure copper. Dispersion-hardening Cu alloys contain

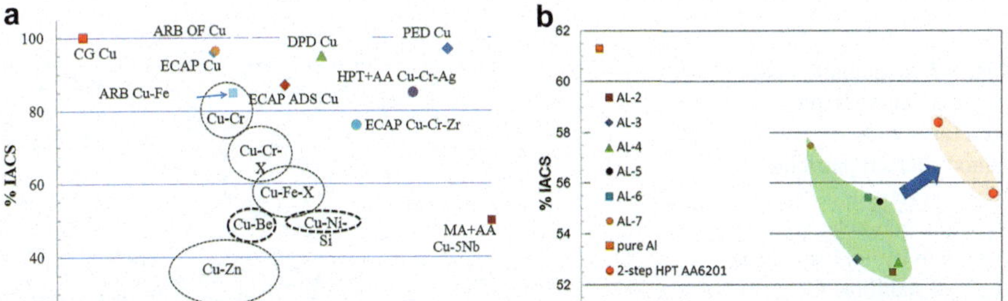

Fig. 3.1 Specific electrical conductivity and ultimate tensile strength at room temperature for coarse-grained and nanostructured states obtained by SPD techniques, pure Cu and its alloys (**a**), pure Al and commercial Al-Mg-Si alloys (**b**). (Reprinted by permission from Springer Nature [11], Copyright 2016)

a small amount of impurity element atoms that are present in solid solution in the quenched state and precipitate out of it as solid-dispersed precipitates during subsequent artificial aging.

The conventional approach used to achieve an optimal combination of enhanced strength and electrical conductivity involves thermomechanical treatments followed by aging. SPD of dispersion-hardening Cu alloys in the quenched state leads to high specific electrical resistivity due to the formation of a nanostructure with small grain size, high dislocation density, and high concentration of atoms of alloying elements in the solid solution. However, the subsequent aging is accompanied by the depletion of the solid solution by atoms of the alloying elements due to the precipitation of particles of secondary phases. Thus, the solid-solution component of the specific electrical resistivity decreases, although the component associated with the precipitated particles of secondary phases slightly increases.

High electrical conductivity can be achieved if the matrix is cleaned from impurity atoms, and the conductivity electrons are not significantly scattered by the precipitates. Precipitation of the particles of small size ~10÷20 nm from the solid solution is the most effective (Fig. 3.2). The alloy in the solid solution state is subjected to SPD followed by aging to obtain such microstructure. The optimal combination of strength, electrical conductivity, and thermal stability can be obtained by selecting SPD and aging parameters that ensure the formation of UFG structure, solid solution decomposition, and the formation of nanodispersed precipitates.

The formation of bulk nanostructured states in dispersion-hardening Cu alloys leads to an increase in strength by about 1.5÷2 times. At the same time, the electrical conductivity remains high and close to 80% IACS (Fig. 3.1).

Al and its alloys obtain less electrical conductivity than pure Cu and its alloys. However, they have lower specific weight and are cheaper than copper materials. As a result, power, car, and aircraft industries have an increased focus on Al and its alloys as electrical materials.

Microstructure refinement of pure Al by SPD provides a multiple increase in the yield strength but does not lead to significant increase in the specific electrical resistivity at room temperature (Fig. 3.1b). By contrast, the specific electrical resistivity

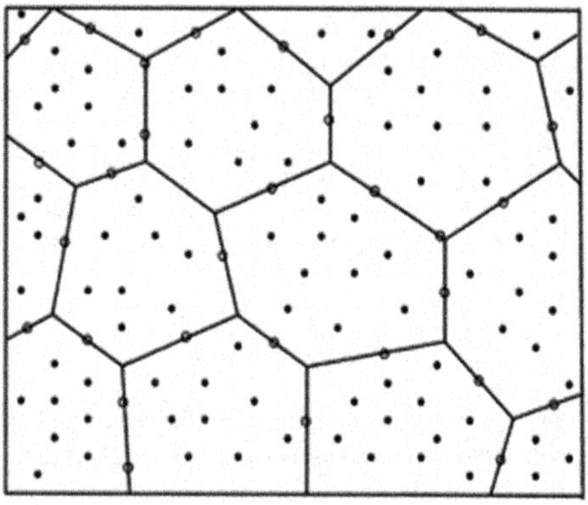

Fig. 3.2 Schematic image of Cu or Al UFG alloy structures with nanodispersed particles, which corresponds to high strength, good electrical conductivity, and enhanced thermal stability. (Reprinted by permission from Springer Nature [11], Copyright 2016)

significantly increases at liquid nitrogen temperature despite the reduced contribution of phonon scattering. This enhancement is mainly due to the contribution of grain boundaries, dislocations, and point defects.

Al alloys of an Al–Mg–Si system (Al alloys of 6XXX series) are the most widely used in the electric power industry for the manufacturing of wires for power transmission lines. Typical values of specific conductivity for these alloys are equal to 86÷92% of pure Al. Simultaneously, the ultimate tensile strength is in the range of 255÷330 MPa (see the area marked in green in Fig. 3.1b). Achievement of such a set of properties is provided by the conventional industrial procedure of solid solution formation, quenching in water, and cold drawing of wires followed by artificial aging.

SPD of Al alloys is usually accompanied by the solid solution decomposition and the precipitation of nanoparticles. This process is often referred to dynamic aging processes and is associated with significantly increased dislocation density and vacancy concentration. As a result, the effective diffusion coefficient is significantly increased by the SPD of Al alloys. Moreover, dynamic aging can lead to a change in the shape of secondary phase precipitations.

One of the new strategies for achieving high strength and good electrical conductivity in Al–Mg–Si alloys is based on this dynamic aging effect. SPD is conducted twice for this effect: first at room temperature, and then at elevated temperatures. As a result, a nanostructured state is formed at the first stage of SPD, and the decomposition of the solid solution with precipitation of nanoparticles of secondary phases occurs at the subsequent stage, which agrees with the scheme in Fig. 3.2.

Alloying of preferentially chemical elements that do not dissolve into the Al matrix is carried out to achieve high electrical conductivity in the development of new Al alloys [11]. Such elements are Cu in AA2024 alloy, Mg in case of AA7075 alloy, as well as Fe, its solubility in Al at room temperature is only 0.025 at.%. These elements are already present in commercial Al alloys and contribute to the formation of dispersed intermetallic particles, ensuring strength and creep resistance. In electrical Al alloys such as AA1350 with a specific electrical conductivity of 62% IACS, the Fe content is usually significant but below 0.8 wt.%. The alloying with Cu and Mg atoms provides not only some hardening but an increase in specific electrical resistivity.

Rare earth (RE) elements such as La and Ce are other promising elements that are immiscible with Al. Adding of up to 10 wt.% of RE elements to the Al matrix results in the following: a significant reduction in grain size and increase of thermal stability up to 250 °C. Thus, SPD of Al alloy with 8.5 wt.% RE elements leads to the formation of a very homogeneous nanostructure with an average grain size of less than 150 nm and $Al_{11}RE_3$ equiaxial intermetallic particles with an average diameter of about 50 nm. These particles are partially dissolved in the Al matrix during the formation of the above-mentioned microstructure via SPD, leading to the formation of supersaturated solid solution. SPD results in a sharp increase in yield stress from 73 MPa (typical for this class of cast materials) to 475 MPa. The obtained UFG structure has maintained excellent thermal stability up to 280 °C for 1 h. This value exceeds the characteristic values of industrial Al alloys demonstrating ~70 °C for Al–Mg–Si alloys and 150–230 °C for Al–Zr alloys used for high-temperature applications.

In recent years, it was discovered that other electrical phenomena and properties are prone to change by the formation of bulk nanostructured states by SPD techniques. Specifically, these properties are giant magnetoresistance, superconductivity, and thermoelectric effects.

3.1.2 Magnetic Properties

As is known, in physics, magnetic substances are subdivided into diamagnets, paramagnets, and ferromagnets [13]. Diamagnetism in materials is caused by a change in the nature of orbital motion of electrons in an atom in an external magnetic field. This phenomenon is typical for any substance. At the same time, the value of magnetic susceptibility χ_m is negative. Paramagnetism is associated with incomplete compensation of spin and/or orbital magnetic moments of electrons in an atom. The value of magnetic susceptibility χ_m is positive for paramagnets and is an order of magnitude higher than the typical absolute value of magnetic susceptibility for diamagnets. Dia- and paramagnetism do not depend on the nature of material microstructure and are extremely weak in physical manifestation, which commonly makes these forms of magnetism of no practical interest.

In ferromagnets, the magnetic susceptibility χ_m is positive and much higher than that of paramagnets. Ferromagnets are characterized by strong exchange interaction between the magnetic moments of neighboring atoms.

Ordered arrangement of the magnetic moments of atoms takes place in certain volumes of a ferromagnet, called domains. Domains are separated from each other by walls. In the walls of domains, the characteristics of neighboring domains change by the orientations of the magnetic moments of atoms.

In ferromagnets, the relationship between the values of the magnetic field H and magnetic induction B is nonlinear, which leads to the effect of magnetic hysteresis during their magnetization reversal. The phenomenon of magnetic hysteresis deals with the dispersion of energy in a ferromagnet, caused by a change in the size of domains and orientation of the magnetic moments of atoms inside the domains when the magnitude and direction of the external magnetic changes.

Several physical values are used to characterize the magnetic properties of a material in a ferromagnetic state. The value of external magnetic field H, in which the ferromagnet has zero magnetization J, i.e., it is completely demagnetized, is called the coercivity H_c. If, in external magnetic field, the magnetic moments of atoms in all domains tend to orient in the direction of external magnetic field, then the ferromagnet is maximally magnetized. The saturation magnetization J_s and the saturation magnetic induction B_s correspond to this state. When external magnetic field H is equal to zero, the ferromagnet can have a remanent magnetization J_r and a remanent magnetic induction B_r.

The linear size of magnetic domains in ferromagnets ranges from thousandths to tenths of a millimeter. In coarse-crystalline materials, each grain can contain several domains.

Heating a ferromagnet to temperature called the Curie temperature and higher destructs the order of magnetic moments of atoms in the domains. As a result, above the Curie temperature, the material is paramagnetic, and below the Curie temperature, the material is ferromagnetic.

Magnetic properties in a ferromagnet crystal depend on its selected crystallographic direction. This phenomenon is called magnetic anisotropy. Magnetic anisotropy arises from the dependence of the interaction of the magnetic moments of atoms on their orientations with respect to the crystallographic axes of the sample. Magnetic anisotropy for a given crystallographic direction in a crystal is calculated as the difference in energies required to magnetize a unit volume of a sample along an easy and a hard axis of magnetization.

There are hard ferromagnetic materials and soft ferromagnetic materials.

Hard ferromagnets have high remanent magnetization, coercivity, and saturation magnetic induction and are used for the manufacture of permanent magnets. The value of the product $(BH)_{max}$, which has the dimension of specific energy, is used to denote the properties of a hard ferromagnetic material. The maximum energy product of high-performance NdFeB permanent magnets can reach the value of 451 kJ/m^3.

The characteristics of hard ferromagnets are wide hysteresis loops. Magnetization reversal faces the problem of high energy consumption. The initial magnetic susceptibility χ_m is small. The coercivity H_c can be increased if the obstacles are created to the movement of magnetic domain walls. By contrast, soft ferromagnets have high initial magnetic susceptibility χ_m and low coercivity H_c. They are characterized by narrow hysteresis loops and low energy losses during magnetization reversal. In this regard, soft ferromagnets are used, for example, for the manufacture of transformer cores. Magnetic susceptibility χ_m and coercivity H_c for soft ferromagnets are more sensitive to changes in the microstructure than that for hard ferromagnets. Particles of non-magnetic phases or pores in magnetic materials impede the movement of the domain walls and increase coercivity. In this regard, soft ferromagnetic materials should not contain such defects.

The first nanocrystalline hard magnetic materials were formed by rapid quenching and annealing of amorphous ribbons. The magnetic properties of such samples did not exceed those characteristic of advanced rare earth materials (Fig. 3.3).

SPD leads to a significant increase in the coercivity H_c of hard magnetic materials (Fe–Pd, R–Fe–B, Fe–Cr–Co, Cu–Co) used as permanent magnets. Such growth relates to the formation of nanostructural states as well as the changes in phase composition. As is seen in Fig. 3.4 (curve 1), the coercivity of the $Pr_{20}Fe_{73.5}B_5Cu_{1.5}$ alloy in the initial state increases with increasing the rotational angle of one of the anvils (strain degree) during HPT at room temperature, where the value reaches a maximum and then decreases. The change in coercivity is associated with the changes in phase composition during HPT. Moreover, annealing of the initial state

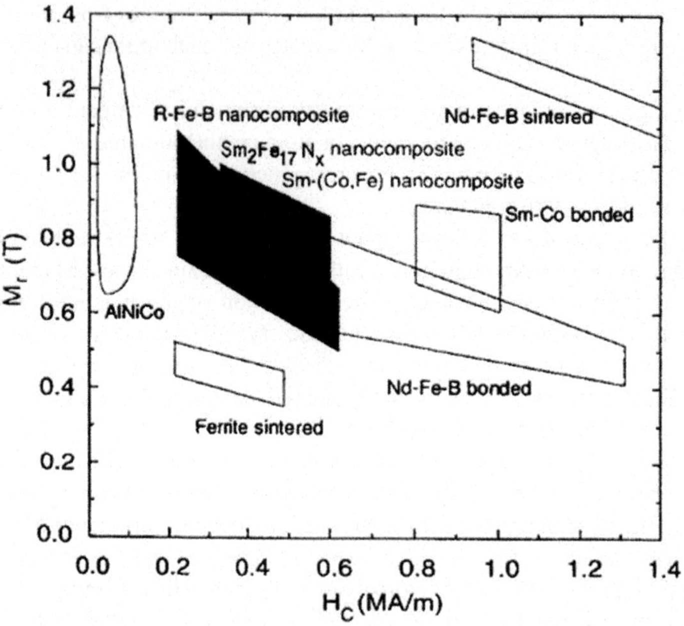

Fig. 3.3 Remanent magnetization versus coercivity for various hard magnetic materials. (Reproduced from [14], Copyright 2003, with permission of John Wiley & Sons)

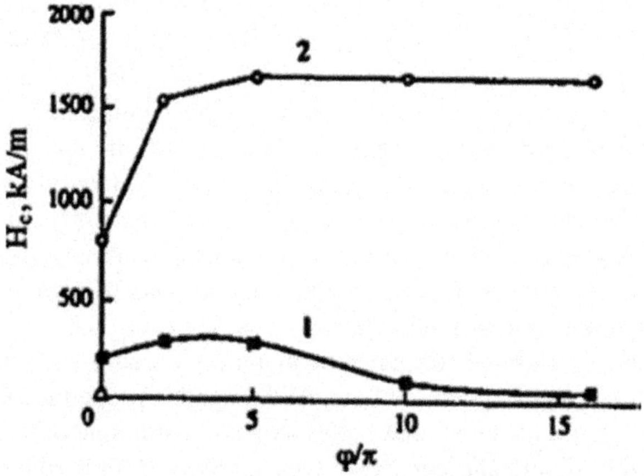

Fig. 3.4 Dependence of coercivity H_c in the $Pr_{20}Fe_{73.5}B_5Cu_{1.5}$ alloy on rotational angle of one of the anvils (shearing strain degree) during HPT under high pressure in a deformed state (1) and deformed state annealed at 600 °C (2). (Reprinted by permission from Springer Nature [15], Copyright 1997)

after various strain degrees leads to restoration of the initial phase composition and increase in the coercivity (Fig. 3.4, curve 2), the latter reaching record values for the alloy $H_c > 1600$ kA/m. This value significantly exceeds the corresponding values typical for the same alloy processed by synthesis or hot deformation.

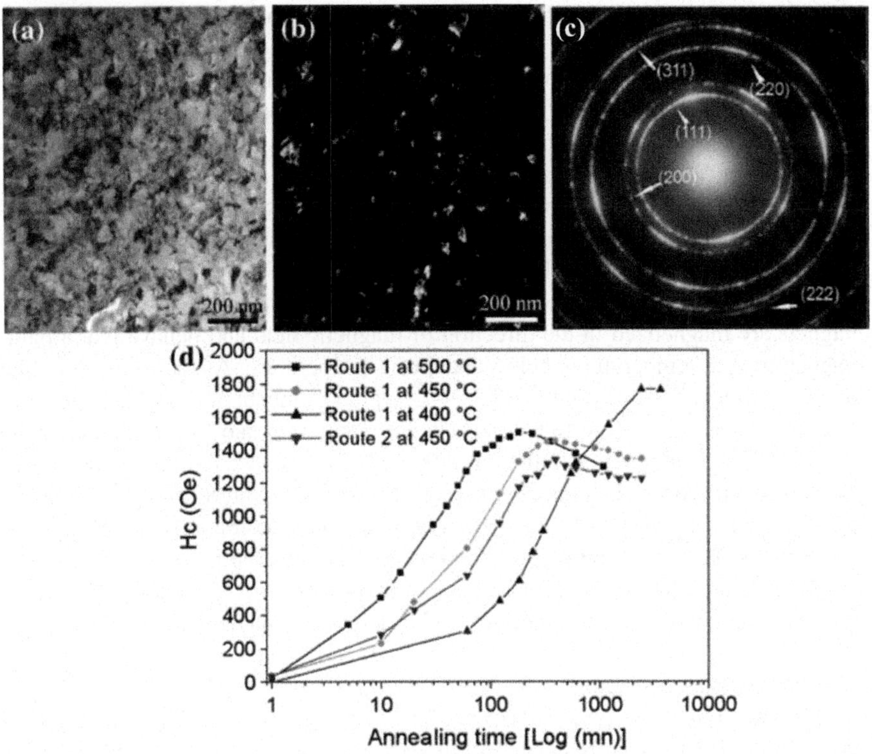

Fig. 3.5 TEM microstructure of FePd alloy after HPT (route 1 – disordered state: bright-field image of a nanoscale microstructure (**a**), dark-field image with several isolated nanosized grains (**b**); diffraction pattern (**c**); and the dependence of coercivity on annealing time at different temperatures for routes 1 and 2 (**d**)). (Reproduced from [16], Copyright 2010, with permission of John Wiley & Sons)

HPT processing of the Fe–Pd alloy both in disordered (route 1) and ordered (route 2) states led to the formation of nanostructured states with ultrafine-grain sizes equal to 50–100 nm (Fig. 3.5) [16]. Moreover, HPT of the ordered state provided significant disordering of the ordered phase L10. At the same time, nanocrystalline grains are clearly visible in the microstructure of the deformed state (Fig. 3.5b). Annealing of the deformed states initiated a first-order phase transformation characterized by the nucleation of numerous ordered regions along grain boundaries. Studies of the effect of annealing temperature and time on coercivity evolution in the disordered (route 1) and ordered (route 2) states showed that the maximum value of coercivity H_c (about 1.8 kOe) is achieved if the disordered state is subjected to HPT with subsequent annealing at a low temperature of 400 °C for 1 hour (Fig. 3.5d).

Formation of nanostructural states in Cu–Co alloys from the as-cast condition through severe plastic deformation by HPT provided a pronounced effect of giant magnetoresistance [16]. The phenomenon of giant magnetoresistance relates to a significant change (by tens of percent) in electrical resistance under the influence of a magnetic field.

Cu–2.2 wt.% Co alloy in the as-cast state had a magnetization equal to only 79 emu/g_{Co} at the insignificant values of $1/H$ [17]. This corresponds to about 50% of the saturation magnetization specifically attributed to Co. The alloy after HPT demonstrated typical superparamagnetic behavior without a tendency to saturation. Superparamagnetic behavior of nanocrystalline ferromagnets relates to the formation of a single-domain state in very small grains. The superparamagnetic state in ferromagnets is characterized by a random orientation of the magnetic moments of domains (grains) in the absence of a magnetic field. In a magnetic field, superparamagnets are magnetized in the direction of magnetic field and behave like a paramagnet even at temperatures below the Curie temperature. At the same time, the magnetization is $J = 18$ emu/g_{Co} at $1/H \rightarrow 0$, which is only about 11% of the saturation magnetization specific to Co. Absence of saturation and low J values for the Cu–2.2 wt.% Co alloy relate to the formation of nanosized Co particles in deformed samples, the size of which is smaller than the limiting size of ferromagnetic domains in Co, which is equal to 4–10 nm. The large fraction of more coarse ferromagnetic Co grains in the Cu–2.2 wt.% Co alloy changes the material behavior. In practice, magnetization of the as-cast alloy reaches saturation at $J_s = 135$ emu/g_{Co}. This value is lower than that of Co. HPT does not lead to saturation magnetization, although its value is approximately three times higher at $1/H \rightarrow 0$ than that of the Cu–2.2 wt.% Co alloy in the as-cast state.

The first ECAP pass on a Cu–6.5 wt.% Co alloy led to a significant increase in the coercivity, which is associated with the interference of the crystal structure defects with the movement of the domain walls [18]. Subsequent ECAP passes via route A resulted in additional increases in the coercivity provided by the refinement of ferromagnetic Co particles down to 10–30 nm by severe shear deformation.

Nanocrystalline soft magnetic materials have the ability to be easily remagnetized and are usually characterized by low coercivity values H_c. The orientation of magnetic moments when changing the direction of the magnetic field can be easily modified and provide small values of coercivity H_c. From the other side, grain boundaries can prevent changes in the orientation of magnetic moments of atoms [19]. It is therefore a classic statement that enhanced soft magnetic properties require very large grain sizes.

Coercivity H_c decreases linearly with increasing grain size in coarse-grained alloys (Fig. 3.6) [20]. By contrast, studies of nanostructured fast-quenched annealed strips made of alloys with different Fe compositions have demonstrated that they are characterized by the reversal tendency (Fig. 3.6). Following the law $H_c \sim D^6$ and reaching the values characteristic of amorphous materials, the coercivity H_c sharply reduces with decreasing grain size D in the range from 100 nm down to 5 nm.

"Finemet" type alloys of Fe–Cu–Nb–Si–B and Fe–Zr–B–Cu systems are the most prominent examples of nanocrystalline soft magnetic materials. Nanocrystalline samples from these alloys are obtained by annealing of fast-quenched amorphous strips. The resulting nanostructures are characterized by grain size from 10 to 20 nm with a two-phase state in the form of small α-Fe crystals arranged in an amorphous

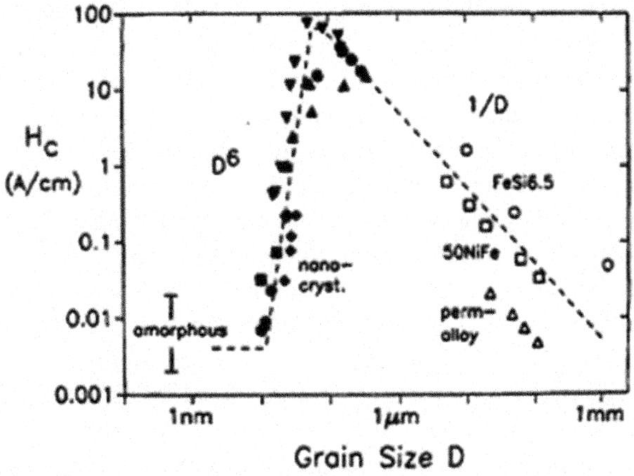

Fig. 3.6 Coercivity dependence on grain size D for different soft magnetic metal alloys: Fe–Ni–Si–B (▲) and Fe–Cu–Nb–Si–B (•), Fe–Cu–V–Si–B (♦), Fe–Zr–B (■), Fe–Co–Zr (▼), Ni–Fe-alloys (Δ, □), FeSi 6.5 wt. % (O). (Reprinted from [20], Copyright 1993, with permission from Elsevier)

matrix. Cu in alloys of the Fe–Cu–Nb–Si–B system enhances the nucleation of α–Fe–Si phase grains, while Nb restrains the grain growth of this phase.

The coercivity enhances as the grain size decreases down to about 0.5 μm due to the difficulty in reorienting the magnetic domains with their walls fixed by the grain boundaries. However, as the grain size decreases below a certain critical size, smaller than about 100 nm, the coercivity values H_c drop sharply. In this case, there is a close correlation between the H_c values and the value of the averaged magnetic anisotropy. Low or negligible magnetic anisotropies are the main conditions for good soft magnetic properties.

The nanocrystalline alloys of the Finemet system obtain a low coercivity H_c up to a temperature of ~300 °C. An increase of the coercivity at higher temperatures is due to the low Curie temperature of the amorphous phase with its value close to 300 °C. At this temperature, magnetic order caused by the exchange correlation in the nanocrystalline α–Fe grains is destroyed, thereby leading to the dramatic changes of saturation magnetization in the nanostructured alloys.

These alloys, having the advantages of large saturation induction, small loss of magnetic order by magnetization reversal, and high conductivity values, are used for the cores of power and instrument current transformers of industrial and high frequency (50–100 kHz) in the manufacture of pulse high-frequency transformers, rotors of electric motors, sensors, switches, and other devices.

Nanocrystalline soft magnetic alloys exhibit excellent magnetic properties such as high magnetic permeability, low coercivity, and relatively high saturation magnetization. Magnetic properties of nanocrystalline soft magnetic materials can be considered within the framework of the so-called random anisotropy model. The magnetic exchange correlation length (domain size) is smaller than the grain size of actual grains. There may be several magnetic domains inside large grains. In nanocrystalline materials, the magnetic exchange correlation length is larger than the grain size. The grain boundaries interfere with the reorientation of magnetic moments in the direction of light magnetization leading to a sharp decrease of the effective anisotropy compared to that characteristic of large grains.

The decrease of the coercivity with the reduction of the grain size below some critical size comparable to that of magnetic domains opens up unique opportunities for a wide application of nanostructured materials as soft magnetic materials.

The coercivity value H_c is greatly influenced by microstructure refinement, an increase of dislocation density, and the formation of a non-equilibrium state of boundaries obtained via SPD of pure soft magnetic metals [19, 20]. Thus, the transition of non-equilibrium grain boundaries into a more equilibrium state during the microstructure recovery via annealing of pure Ni subjected to SPD leads to a 40% decrease of the coercivity value. At the same time, the grain size remains nearly unchanged. Increasing the annealing temperature leads to grain growth and the subsequent coercivity value decreases. Behavior of the coercivity value is characteristic of nanostructured Co obtained by high pressure torsion. In this case, the grain size after SPD processing was about 100 nm with the size of the magnetic domains significantly exceeding the grain size. However, the non-equilibrium state of grain boundaries prevented the magnetization reversal.

Nanostructured design using SPD techniques and possibly subsequent annealing can contribute to a significant increase in the magnetic properties of both hard magnetic and soft magnetic materials. The comparability of the grain size and the characteristic magnetic exchange correlation length results in the enhancement of the magnetic properties of nanostructured materials.

3.2 Chemical and Biomedical Properties of BNM Materials

3.2.1 Corrosion Properties

Items made of various alloys are exposed to aggressive environmental influences during their applications, resulting in corrosion processes. Corrosion is a spontaneous destruction of metallic materials due to chemical or physical–chemical reactions with the environment. At present, the total amount of direct corrosion losses at the lowest estimate is 3–4% of the gross national product of various countries. Increasing the corrosion resistance of materials is therefore a highly topical problem.

Metallic materials are more or less prone to corrosion. Standard electrode potential of the metal is a quantitative characteristic of the tendency of a metal or its ion to certain reactions. The value of standard electrode potential quantifies the reduction potential of metal and the oxidation potential of its ions. For example, the standard electrode potentials are −2.37 V for magnesium, −1.63 V for titanium, −0.76 V for zinc, −0.44 V for iron, and + 0.34 V for copper. By the values of reduction potential, it is possible to estimate the rate of spontaneous flow of oxidation-reduction reactions in a certain direction.

Under standard conditions, standard electrode potential is defined as a difference in electrical potentials between an electrode and being in contact with its electrolyte (typically between a metal and an electrolyte solution). The appearance of the electrode potential is due to the charged particles transport through the phase boundary, double electric layer, specific adsorption of ions or orientation adsorption of polar molecules (including solvent molecules).

The influence of grain size on corrosion resistance of metals and alloys is of different character [21]. Certain difficulties are caused by the establishment of such interrelation. These difficulties are determined by the changes in a number of microstructural parameters in the course of various thermomechanical effects aimed at formation of microstructures with different grain sizes. These parameters, as well as the grain size, can influence corrosion resistance. Among such parameters are: internal stresses, defect density, segregation of alloying element atoms at grain boundaries, etc. The microstructure evolution during heat treatment, following the procedure of forming a microstructure with a defined grain size, makes it even more difficult to find the desired relation.

The authors of recent reviews [21, 22] state that at present there is no consensus on the improvement or deterioration of corrosion properties as a result of the formation of bulk nanostructured states by SPD techniques. Apparently, it is associated with a variety and peculiarities of microstructures formed at different stages and using various SPD techniques.

The general regularities and mechanisms of corrosion, depending on the degree of the formation of UFG materials through SPD, are presented in the review [22]. Four regularities of the corrosion resistance changes with increasing degree of deformation are singled out (Fig. 3.7). Specifically, the term "corrosion resistance"

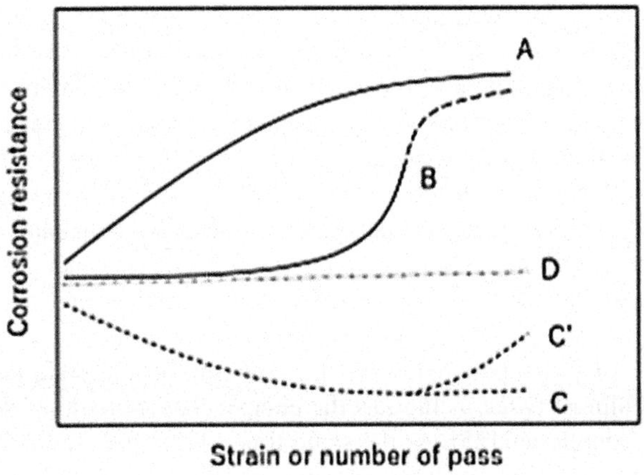

Fig. 3.7 Classification of regularities of corrosion resistance change with increase of plastic deformation degree. (Reproduced from [22], Copyright 2019, with permission of The Japanese Institute of Metals and Materials)

means one of the following physical values: corrosion potential, polarization resistance (the ratio of the potential increment taken in the small vicinity of the equilibrium potential to the current increment), corrosion current, mass loss.

Regularity A is characteristic for a wide range of materials, corrosion resistance of which in passivative media increases with decreasing speed and tends to a constant value with increasing degree of deformation. Thus, the surface of metal enters into an inactive state as a result of thin oxide or salt film formation. This trend seems to be consistent with the nature of the transformation of stored free energy as the deformation degree changes. As of regularity B, an increase in the degree of deformation leads to corrosion resistance improvement with a sharp increase in the intermediate stage of the SPD. Such regularities may be related to some specific changes in the microstructure or to the formation of the UFG structure. In case of regularity C, the corrosion rate increases (corrosion resistance decreases) along with the defect density during SPD process. Finally, the C′ -type regularity is associated with a decrease in corrosion resistance at early stages of SPD, but with its increase in the formation of the SPD structure, corrosion resistance increases as well.

Grain size is a key parameter in determining the corrosion resistance. For materials subjected to different methods of microstructure refinement, including SPD techniques, the Hall–Petch relationship linking the corrosion current and grain size has been established within the limits of error [23].

Enhancement of the deformation degree leads to the decrease of grain size and, therefore, to the increase in the length of grain boundaries as well as dislocation density. As a result, the stored free energy is accumulated with the simultaneous growth of the electrode potential, which is, according to the laws of thermodynamics, directly proportional to the stored free energy. However, estimates exemplified by pure copper subjected to severe cold deformation indicate slight changes in the electrode potential (0.1 mV) even when a very high dislocation density is reached (10^{15} m^{-2}). At the same time, experimental studies point to significant changes in the corrosion potential, which typically is not equal to either equilibrium or standard electrode potentials, by the value of 30 mV in 3.5% NaCl and even 100 mV in the Livingstone etching solution. This testifies to the necessity to consider the contribution of the electrode potential to the value of the corrosion potential as well as the accelerated kinetics of reactions, changing due to the high potential of grain boundaries and dislocations and the associated stored local free energy.

Estimation of potential difference at grain boundaries, based on the value of additional free energy (7×10^{-21} J) per 1 atom in the grain boundary, amounts to the value equal to −22 mV. However, a minimum potential difference of 30 to 40 mV is required to start galvanic corrosion between the grain boundary and the grain body [24].

The energy of grain boundaries in the non-equilibrium state caused by SPD (~1.2 J/m^2) is almost twice as high as the energy characteristic of the equilibrium state of grain boundaries [25]. At the same time, the potential difference between the non-equilibrium boundary and the grain body reaches 44 mV. This value is sufficient for the formation of local nanoscale corrosive galvanic couples as well as for the electrochemical processes. The formation of corrosive galvanic couples is also possible between the nuclei of individual dislocations and the grain body. Both theoretical and experimental assessments point to a local potential difference of several dozens of millivolts.

Figure 3.8 demonstrates the influence of the parameters of the microstructure formed at different stages of SPD on corrosion process [22].

The initial state at stage 1 is characterized by the absence of dislocations in the recrystallized grains body and uniform distribution of electrode potential. A small local reduction of the potential occurs at grain boundaries which involve very limited volume in a material. Stage 2 is characterized by numerous local potential decreases at the dislocation nuclei and grain boundaries. In this case, the reduction of the total potential corresponds to the growth of stored free energy. Formation of perfect UFG structure at stage 3 is characterized by the potential reduction only at grain boundaries due to the almost full absence of defects in the grain body. The value of the total potential reduction has increased, as the grain boundaries obtain a higher free energy. Subsequently, the potential in the grain body returns to the initial level due to absorption of lattice dislocations by grain boundaries. Annealing of UFG structures obtained by SPD techniques is accompanied by recovery processes, small grain growth, and transition of grain boundaries to a more equilibrium state (stage 4). Thus, reduction of the local potential occurs only at the equilibrium grain boundaries and in less expressed manner.

The corroded surface relief correlates with local potential reduction and oxidation ability of the solution characterized by the equilibrium electrode potential of the E_c cathode reaction. If E_c is bigger than both the potential of the body of grains $E_{a.g}$ and the potential of grains boundaries $E_{a.gb}$, as in the case of aggressive media, general corrosion may occur. In this case, the corrosion kinetics both in dissolution and passivation is determined by the total stored energy level. If, however, the equilibrium potential E_c of the cathodic reaction is between $E_{a.g.}$ and $E_{a.gb}$, corrosion becomes selective and occurs on those defects whose corrosion potential is less than that of the E_c. The corrosion rate becomes sensitive to the microstructure evolution and increases or decreases during the formation of the UFG structure depending on the E_c level.

A very high density of grain boundaries, contributing to the formation of closely located galvanic couples of anode–cathode reactions, can serve as one of the reasons for an increase of corrosion resistance by the introduction of UFG structure. As a result, the corroded surface should be more homogeneous. Another possible reason is restriction of galvanic couples' sizes by grain size that should suppress reactions in weak or diluted electrolyte solutions.

In general, issues of corrosion of UFG metal materials are very complicated and require detailed study in each case.

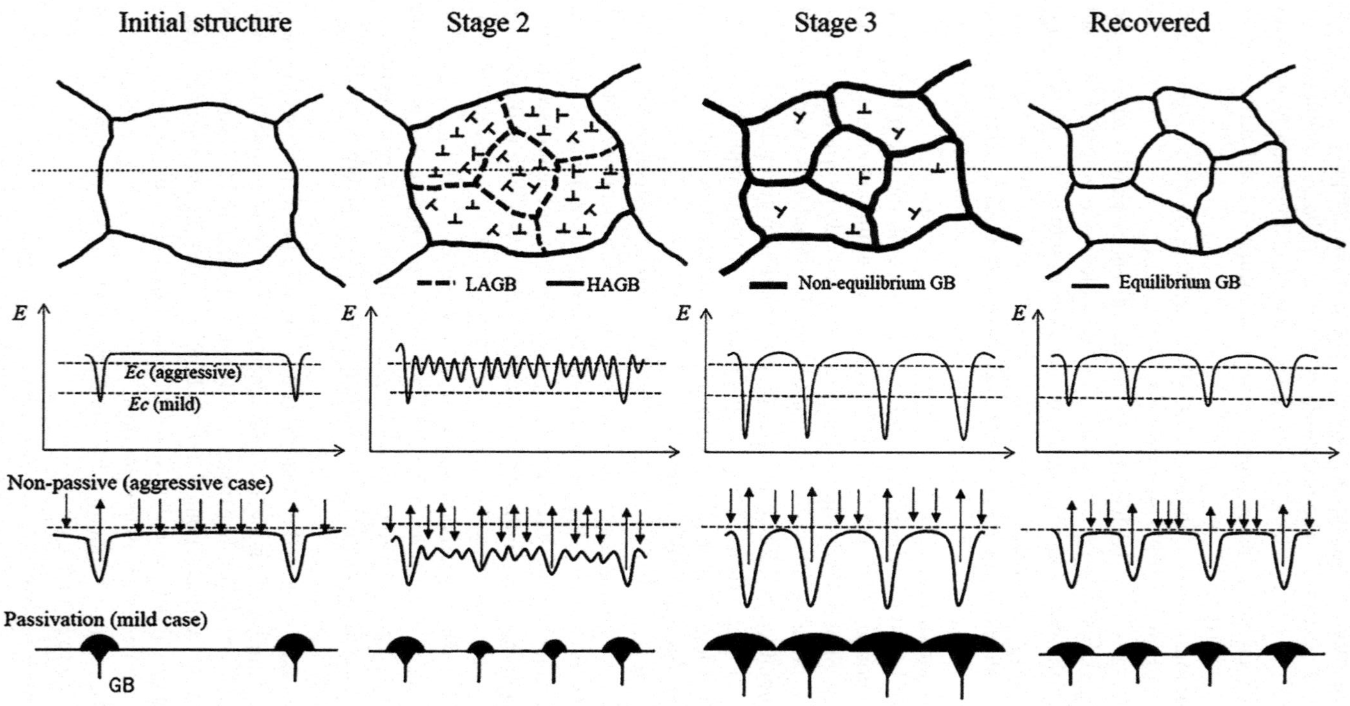

Fig. 3.8 Schematic diagram demonstrating microstructures at different stages of SPD (top), corresponding distribution of anode reaction potential (metal oxidation) (center), as well as corroded surfaces without passivation in relatively aggressive (large E_c) solutions and passivations in weak electrolyte (small E_c) solutions (bottom). E_c – equilibrium potential of cathode (reduction) reaction. (Reproduced from [22], Copyright 2019, with permission of The Japanese Institute of Metals and Materials)

3.2.2 Biomedical Properties

Increased structural and functional properties (elasticity, fatigue strength, fracture resistance, the ability to take and maintain a certain geometric shape, corrosion resistance, biological neutrality, and other) are the factors that determine a wide application of products made of metallic materials in medicine. Such products include bone implants and anchors, surgical instrument sets, stents for coronary vessels (metal cylinder-shaped struts used to expand narrow spaces in the artery with cholesterol deposits), suture material for vascular and neurosurgery, endoprostheses (internal devices replacing an organ function) in ophthalmology and dentistry, and many other medical products [26].

The corresponding paragraphs of this textbook describe nanostructuring by SPD techniques enabling the formation of advanced properties, thus making the use of UFG metallic materials as biomaterials rather attractive proposition [26, 27]. Specifically, biocompatibility is the most important property for metallic biomaterials for medical application [28].

"Biocompatibility" is a term describing the ability of biomaterials to perform certain functions with an appropriate host response. The composition and physicochemical properties of biomaterials should remain stable throughout their lifespan. Biocompatibility is determined by the nature of the interaction between biomaterials and human body tissues, i.e., by the processes occurring on their adjacent surfaces. The interaction between proteins as well as other physiological macromolecules and the surface of biomaterials can initiate inflammatory and/or immune responses, followed by the establishment of a stable equilibrium resulting from repair or regeneration processes. At the same time, the inclination to rejection, the rate of corrosion or degradation of biomaterials, the impact on homeostasis (relative dynamic constancy of the internal environment and stability of basic physiological functions) should be minimal [29].

Pure Ti is one of the main metallic biomaterials in medical application [29]. Its highest biocompatibility compared to other metals is due to the formation of a 50÷100 Å thick oxide film protecting the surface of pure Ti. A mineralized bone matrix (intercellular substance of bone tissue with a high concentration of calcium salts) covered by proteins of blood plasma, such as cryoglobulin and vitronectin is formed on the film. The complete biological indifference of the oxide film ensures the lack of bone tissue response to the metal as a foreign body and recognizes it immunologically to its kind.

The formation of the UFG structure in pure Ti provides an increase in histocompatibility (degree of similarity of two tissues) in the range from 30% to 20fold values. The rate of proliferation of preosteoblasts (cells differentiating [transforming] into osteoblasts, i.e., young bone tissue cells with a diameter of 15÷20 microns synthesizing mouse MC3T3-E1 matrix) increases with the enhancement of the average length of grain boundaries on the implant surface (Figs. 3.9 and 3.10) [30].

Biocompatibility is strongly influenced by the surface condition of Ti implants [30]. Surface polishing increases the rate of cell attachment and multiplication (Fig. 3.9) [30], and the effect is much stronger for UFG Ti rather than its CG counterpart.

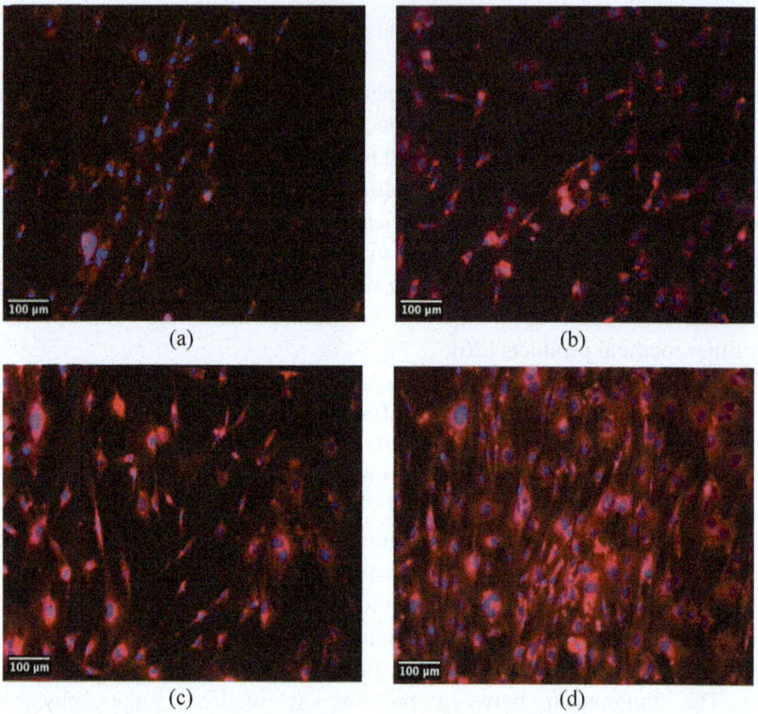

Fig. 3.9 Fluorescent images of mouse preosteoblasts after 72 h for (**a**) CG-Ti unpolished, (**b**) CG-Ti polished, (**c**) UFG-Ti unpolished, and (**d**) UFG-Ti polished [30]

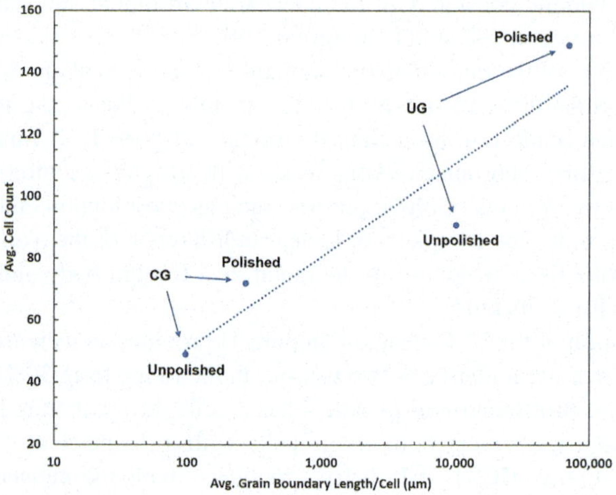

Fig. 3.10 Attached number of MC3T3-E1 cells after 72 h of seeding depending on the average length of grain boundaries per attached cell [30]

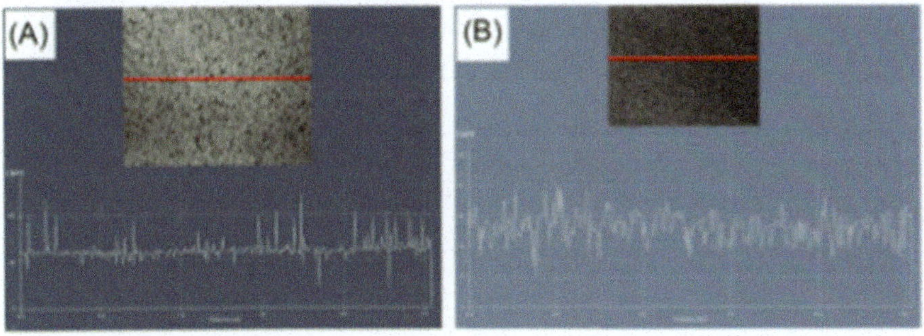

Fig. 3.11 General view of the surface and profile in the section (red line) of the CG (**a**) and UFG (**b**) Grade 4 Ti samples after etching in the mixture of acids for 20 min. (Laser Scanning Microscopy). (Reprinted from [31], Copyright 2018, with permission from Elsevier)

Surface modification by etching or deposition of bioactive coatings also increases the biocompatibility of the implants [31]. Acid etching removes the impurities from mechanical processing on the surface of UFG Ti implants as well as polishes it during finishing treatment.

The reaction between hydrofluoric acid and the TiO_2 dioxide film during pretreatment ensures the formation of a thin film of fluorides and hydrides on the surface of implants for maxillofacial surgery. However, as for osteosynthesis (setting a broken bone), such processing may result in the fatigue resistance decrease due to the formation of a deep surface relief [26]. The surface relief created by acid etching is deeper for UFG Ti (Figs. 3.11, 3.12 and 3.13) [31]. The deep relief provides an increased rate of adhesion of fibroblast and preosteoblast cells.

The emission of Ti ions into the human body from UFG Ti implants is extremely low and much less than the histological level (level of presence of dangerous cells and neoplasms) established for Ti and equal to 5 ppm. By contrast, concentrations of Ti ions in the tissues surrounding CG Ti implants with grain size ≤1000 μm are usually much higher than in blood or saliva.

The deposition of biocompatible coatings on the surface of Ti implants and Ti-based alloys considerably increases the integration of implants and bone tissue [33–35]. At the same time, the actual problem is the synthesis of biocompatible coatings made from inorganic (Ca and P containing) and organic (containing bioactive and bioinert molecules) components on the implants. Biocompatible coatings contain hydroxyapatite (HA), which is the main crystalline component of human bone and a drug stimulant for osteosynthesis, but they obtain low adhesion to the Ti substrate as well as low mechanical strength. Plasma electrolytic oxidation (PEO) has recently been widely used for deposition of coatings containing Ca and P [33]. Application of this technique accompanied by high voltages contributes to microdischarges in the coating (Fig. 3.14a), its recrystallization, and intensive growth. The stable titanium dioxide modifications (rutile and anatase) are contained in the coatings. The deposition method featured by electrochemical oxidation and numerous melting and hardening of the microdischarged areas provides strong coupling to the surface. This method makes it possible to obtain the coatings with adjustable porosity and pore size from 0.1 to 10 μm (Fig. 3.14b) [34].

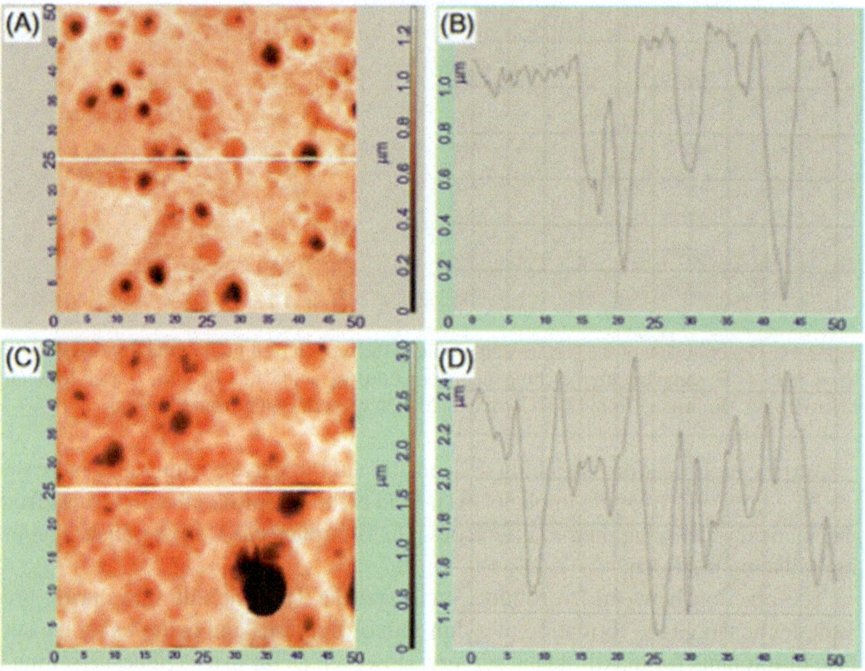

Fig. 3.12 Typical images of the surface topography of samples (scanning area 50 × 50 μm) (**a**, **c**) and profiles (**b**, **d**) of CG and UFG Grade 4 Ti after etching. (**a**, **b**) Surface of CG Ti (c, d) surface of nano-Ti. Atomic Force Microscopy. (Reprinted from [31], Copyright 2018, with permission from Elsevier)

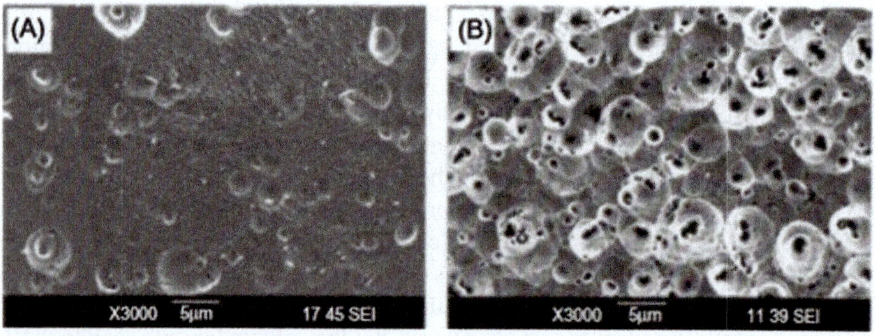

Fig. 3.13 Surface of CG and NS samples of Grade 4 Ti after mechanical polishing and etching in the mixture of acids 30% $HNO_{ck.yt}$ + 3% HF + H_2O for 20 min: (**a**) CG Ti surface; (**b**) nano-Ti surface. SEM image. (Reprinted from [31], Copyright 2018, with permission from Elsevier)

The coating elastic modulus gradually changes from the metal implant to the bone, contributing to improved biocompatibility. High porosity of the PEO coating provides a better attachment of the osteoblast to the implant surface.

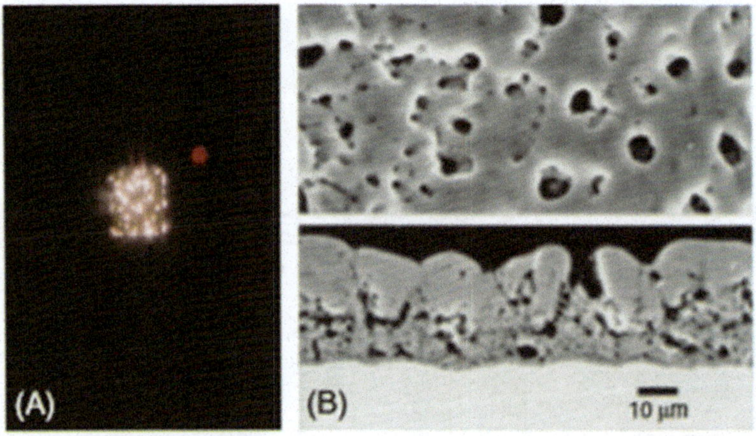

Fig. 3.14 Images of microdischarges on the sample surface during the PEO process (**a**); SEM image of PEO Ca- and P-containing coating on titanium, top view (above) and cross-section (below) (**b**). (Reprinted from [31], Copyright 2018, with permission from Elsevier)

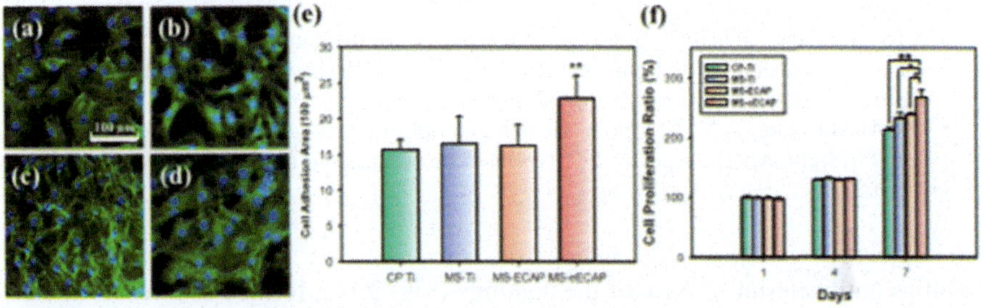

Fig. 3.15 Pre-osteoblast cells spreading on the surfaces and proliferation rate. Fluorescence images of F-actin (green) and nucleus (blue) stained cells on the (**a**) CP Ti, (**b**) MS Ti, (**c**) MS-ECAP Ti, and (**d**) MS-eECAP substrates after 3 days. (**e**) The average cell adhesion area on each substrate by measuring the region of F-actin fluorescence using J software. The fluorescence images of F-actin used false color from red to green, and merged with the nucleus fluorescence images. (**f**) Proliferation rate of the pre-osteoblast cells with a reference of the CP Ti substrate. (Reprinted by permission from Springer Nature [36, 37], Copyright 2018)

The coating should produce an antibacterial effect on the implant media. Polymethimethocrylate (PMMA) coatings based on protein-resistant polyethylene glycol (PEG) hydrophilic polymethacrylic acid deposited on the Ti surface have a bactericidal effect, but they degrade within 8÷12 days [36].

Antibacterial activity is also characteristic of water medium stable molybdenum disulfide MoS_2 coatings. Low cytotoxicity and genotoxicity of these coatings improve cell attachment, thus eliminating anticoagulation, restonosis (reconstriction in the vascular area where angioplasty has already been performed or a stent has been placed), and thrombosis. This is evidenced, in particular, by the results of the comparative analysis of the implant surfaces images with the pre-osteoblast cells, diagrams reflecting the size of the areas and their growth rates (Fig. 3.15) as

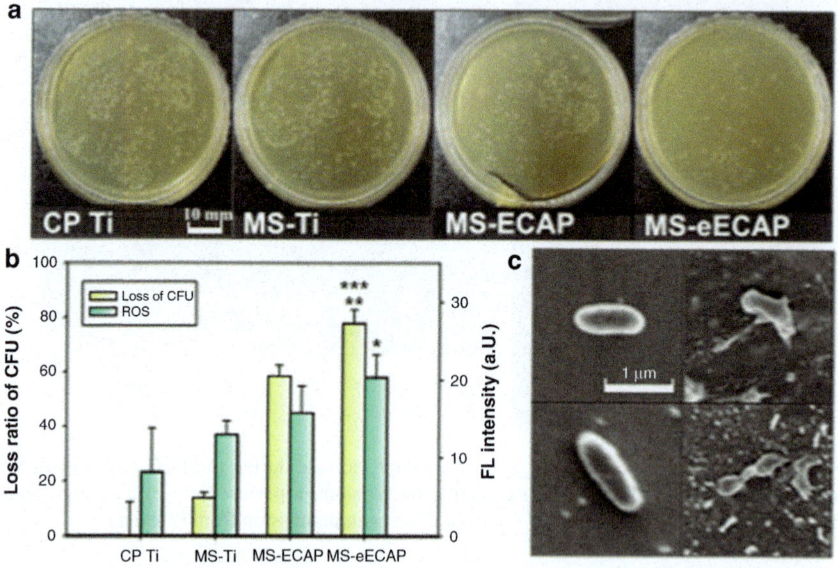

Fig. 3.16 Antibacterial activity of MoS_2 coated specimens. (**a**) Photographs of *E. coli* growth on bacterial culture medium (overnight incubation at 37 °C) and (**b**) bacterial growth inhabitation rate of each substrate at 37 °C for 6 h. (**c**) SEM images of *E. coli* exposed to the CP Ti and the MS-eECAP substrates at 37 °C for 6 h. The left and right side images *E. coli* of the surface of the CP Ti and MS-eECAP Ti, respectively. (Reprinted by permission from Springer Nature [36, 37], Copyright 2018)

well as antibacterial activity of the coatings (Fig. 3.16), formed as a result of electrostatic deposition of molybdenum disulfide MoS_2 nanoflakes sized less than 50 nm on substrates of non-etched CG Ti (MS), non-etched UFG Ti (MS-ECAP), etched UFG Ti (MS-eECAP) [36, 37].

Thus, pure UFG Ti with high structural properties and biocompatibility may well replace Ti-based alloys as a biomaterial. Moreover, unlike Ti alloys, pure Ti does not contain harmful impurities that have adverse effects on the human body and limit the scope of Ti medical application.

Nevertheless, the use of SPD techniques to improve the biomedical properties of Ti alloys seems promising [27]. Wide medical application of two-phase ($\alpha + \beta$) titanium alloys, such as Ti–6Al–4 V and Ti–6Al–7Nb, is due to their inherent excellent mechanical properties and sufficient biocompatibility. This is also facilitated by Young's modulus, which is reduced compared to that of pure Ti and is close to Young's modulus of bone, resulting in the reduced stresses between the loaded bone and the implant. These alloys contain toxic V and less toxic Al. The Ti–6Al–7Nb alloy, which does not contain harmful V, has an advantage in this regard. Earlier, in Chap. 2, the positive effect of SPD on increasing the strength properties of these alloys was demonstrated.

The use of SPD techniques also seems promising for increasing the strength properties of Ti–Nb and Ti–Mo alloys that contain neither V nor Al, but have a lower Young's modulus than the Ti–6Al–4 V alloy.

The same is true for β-Ti alloys containing such non-toxic alloying elements as Nb, Mo, Zr, and Ta as well as no toxic impurities. β-Ti alloys, due to their high β-phase content, have a low Young's modulus of 55–90 GPa. The low Young's modulus value in these alloys is observed in the single-phase solid solution state, which is characterized by low strength. Aging results in precipitation of ω and α phase particles and increased strength, but at the same time, Young's modulus values also enhance. Thus, SPD techniques are applied to obtain simultaneous high strength and low elasticity in these alloys.

HPT processing of β-Ti Ti–15Mo alloy certified for medical applications resulted in a significant refinement of the single-phase solid-solution microstructure down to grain sizes less than 100 nm. However, a deformation-induced ω phase was formed, which provided a sharp increase in the elastic modulus. Post-deformation aging was accompanied by the formation of a two-phase microstructure consisting of grains of α and β phases, which was also characterized by an increased value of the modulus of elasticity.

SPD processing of a Ti–Nb–Ta–Zr alloy system that is less prone to the ω phase formation proved quite efficient. Thus, the Ti–29Nb–13Ta–5Zr alloy subjected to HPT demonstrated microstructure refinement and increased in the flow stress from 550 to 800 MPa with a constant elastic modulus. ECAP of the biomedical Ti–35Nb–6Ta–7Zr alloy was accompanied by significant microstructure refinement as well.

The refinement of β-Ti alloys' microstructure can also result from multiple twinning and/or martensitic transformation β → α". Nanostructured β-Ti alloys are characterized by excellent in vivo biocompatibility, as evidenced by increased attachment and multiplication of cells.

Shape memory effect, superelasticity, good biocompatibility, and corrosion resistance in vivo of Ni–Ti alloys with 54–60 wt. % Ni are also major factors for their widespread use in medicine. Elastic modulus of these alloys in the martensitic state is less than half of elastic modulus of pure Ti, which reduces the load on the bone tissue around implants. SPD increases strength, reactive deformation, and stress, and brings the range of superelasticity manifestation to human body temperature.

Nanostructuring leads to an accelerated multiplication of cells on the surfaces of stainless steels, boosts metabolism (the totality of an organism's chemical reactions), and increases characteristic level (expression level) of proteins (fibronectin, actin, vinculin) and osteoblast parameters (cell area, perimeter, Feret diameter) [38].

Nanostructuring is also a promising tool for self-absorbable vascular stents or bone implants widely used in medicine [39–41]—as is the case with at least a five-fold improvement in the corrosion resistance of the UFG biodegradable Mg–1Ca alloy compared to its CG counterpart.

SPD makes it possible to control the biodegradation processes of Mg and its alloys, improves surface morphology [41], and allows solving issues related to excessively high rates of degradation of magnesium products and hydrogen evolution during corrosion in the human body. As is the case with other metallic materials, nanoroughness may be the main determining factor in cell attachment and proliferation to the surface. Other possible reasons include hydrophilic nature, surface chemistry, and texture.

Thus, SPD-induced nanostructuring, along with improved structural properties, leads to an increase in biocompatibility of a number of metals and alloys used in medicine and, among all of them, the primary role is assigned to pure Ti [27].

3.3 Shape Memory Effects and Superelasticity

The last decades are characterized by active development and use of alloys with shape memory effect to manufacture products for medicine, electrical engineering and electronics, aviation and space, pipeline transport, automotive industry, etc.

The most prominent representative of such alloys are alloys based on titanium nickelide (TiNi) near the equiatomic composition (50:50 at. %). Products made of these alloys possess high strength and ductility, significant thermomechanical memory effects, including one-time and reversible shape memory, superelasticity with high mechanothermal and thermocyclic durability, corrosion resistance, biocompatibility, etc. In addition to the above, these alloys generally have a relatively simple chemical composition. In this regard, despite their relatively high cost, alloys of the TiNi system are considered to be functional materials of the new generation [42, 43].

The shape recovery or memory effect in the intermetallic compound TiNi was discovered in the early 1960s of the twentieth century [44]. This effect consists in the recovery of the initial shape during heating a sample and that shape has been changed as a result of deformation at a lower temperature (in the martensitic state). The nature of this effect is as follows: in the initial high-temperature austenitic state, the intermetallide has an atomically ordered B2 structure with a cubic lattice (of the CsCl type) with a lattice period $a_{B2} \approx 0.301$ nm. The austenitic phase B2 transforms into the martensitic phase B19' during cooling. B19' has a monoclinic (distorted orthorhombic) unit cell with parameters (at room temperature): $a \approx 0.290$ nm, $b \approx 0.411$ nm, $c \approx 0.465$ nm, $\beta \approx 97°$. This results in the lattice deformation, which is described by the transformation of a face-centered tetragonal cell with parameters $a_{B2} \approx 0.301$ nm, $b_{B2} = c_{B2} \approx 0.426$ nm in the lattice B2 into a unit cell B19'. It means that the B19' phase has a unit cell in which the interatomic distances along the a and b axes are smaller than a_{B2} and b_{B2}, respectively, while the c axis is considerably larger than the c_{B2} axis. The maximum lattice strain at B2 → B19' transformation is ≈11% and determines the theoretical (crystallographic) resource of reversible strain, that is, the maximum strain that can be "reversed" during the B19' → B2 transformation. During "free" cooling, the TiNi alloy sample does not change shape because the anisotropic stresses arising from the transition from the cubic B2 lattice to the monoclinic B19' lattice are compensated by the equally probable realization of 24 orientational martensite lattice variants. When heated, the inverse B19' → B2

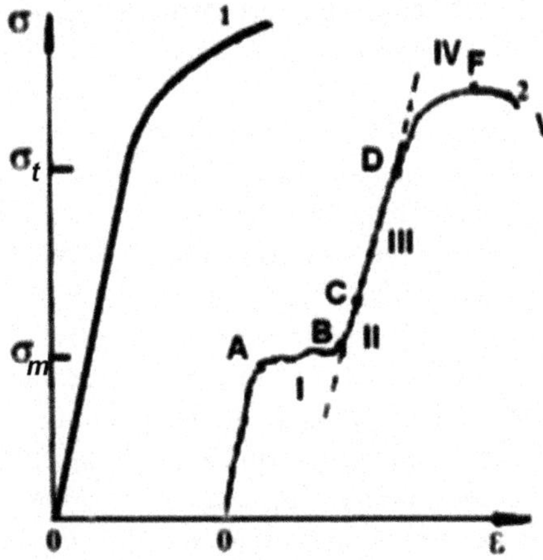

Fig. 3.17 Tensile diagrams for TiNi-based alloys in different structural states: on the left is the curve characteristic of the structure with stable austenite B2 at ($T_d \geq M_d$), on the right is the curve characteristic of the structure with deformation-induced martensite ($T_d < M_s^\sigma$). (Reprinted by permission from Springer Nature [42], Copyright 2008)

transition occurs. The direct transition from the austenitic state to the martensitic state during cooling is characterized by the start (M_s) and finishing (M_f) temperatures of the martensitic transformation. The reverse transition from the martensitic state to the austenitic state during heating is characterized by the start temperature (A_s) and finishing temperature (A_f) of the reverse transformation. Specific values of the characteristic transformation temperatures depend on the chemical composition of alloys, phase state, and structural parameters of TiNi alloys.

Martensite in intermetallic TiNi alloys may form as a result of temperature reduction, as well as under the action of mechanical stresses. At deformation temperatures $T_d \geq M_d$, where M_d is the temperature above which the strain-induced martensitic transformation is suppressed, the behavior of TiNi alloys does not differ from that characteristic of other metallic materials (Fig. 3.17, left curve). When deformed in the temperature range $M_s^\sigma < T_d < M_d$ "deformation martensite" occurs under the stresses developed during plastic deformation of B2-austenite. At deformation in the interval $M_s < T_d < M_s^\sigma$ before the beginning of plastic deformation, "stress martensite" is formed, which corresponds to the appearance of the so-called "phase" yield stress σ_m (Fig. 3.17, right curve) at a stress much lower than the usual ("dislocation") yield stress σ_t.

If the alloy of TiNi system is strained at temperatures below M_f, (in the martensitic state), when a small stress σ_m is reached (Fig. 3.17), the deformation is carried out due to "martensite reorientation" when the martensite crystals or components of their internal twins favorably oriented in relation to the direction of the acting load grow, absorbing unfavorably oriented ones, which eventually results in deformation at the point A to B in Fig. 3.17. This strain due to martensite reorientation can in the limit reach ~11% as in the case of stress martensite formation. This deformation remains if the load is removed at temperatures below A_s. But if the temperature is raised above A_s and A_f, the reverse transition from the B19' to B2 phase occurs and all deformation induced by "martensite reorientation" is cancelled, resulting in the product shape recovery that was before the deformation in the martensitic state. This *is a shape memory effect induced by martensite reorientation.*

If the TiNi alloy is deformed at a temperature in the range $M_s < T_d < A_s$, when σ_m is reached (Fig. 3.17), the deformation is carried out due to "austenite→martensite stressed transformation." The crystal lattices of the obtained martensite are also oriented favorably in relation to the direction of the acting load, which gives a deformation (in the area from points A to B in Fig. 3.17) up to 11%. If the load is removed at a temperature above A_f, this deformation induced by martensitic transformation returns, the product restores its original shape, i.e., the effect of superelasticity (pseudoelasticity) is realized, while the reversible deformation is limited to the same value: ~11%. If the load is removed at a temperature below A_s, the induced deformation will remain, and subsequent heating through the A_s-A_f interval will lead to a reverse martensitic transformation and return of the deformation. *This is the shape memory effect induced by the martensitic transformation.* Heating after deformation through the M_s-M_f interval realizes the shape memory effect induced by both the martensitic transformation and the reorientation of martensite. In the case of deformation in the A_s-A_f temperature interval, a partial shape memory effect can be implemented by subsequent heating above A_f and a partial superelastic strain reversal due to martensite reorientation during unloading.

If TiNi specimens are loaded with a load exceeding σ_t (point D on the curve in Fig. 3.17), the strain occurs due to the ordinary flow of dislocations and the shape memory is not implemented.

The thermoelastic martensitic transformation leads to the following effects:

1. Shape memory effect (SME) is the ability of the alloy to partially or completely return the strain acquired earlier, in the martensitic state, and, accordingly, to recover the original shape.
2. The effect of generation of reactive stresses, which is implemented during heating a pre-deformed sample under conditions of external mechanical resistance.
3. The effect of superelasticity (or pseudo-elasticity) is the ability of an alloy to experience a significant amount of inelastic strain during loading, which returns fully or partially when the load is removed.
4. Large plasticity effect is the ability of an alloy to undergo significant (up to 100%) strain under loading in the temperature range M_s–M_d due to the addition of the deformation of martensitic transformation, and large uniform plastic deformation resulting from inclination of martensite to strain hardening.

The formation of martensite is significantly affected by grain size and the state of the initial austenite [43, 45]. It is possible to form UFG states with different grain sizes in TiNi alloys using SPD techniques. SPD by high pressure torsion of these alloys leads to the formation of an amorphous state [46]. Subsequent annealing allows the formation of microstructures with grain size of a few tens of nanometers. The $Ti_{49.4}Ni_{50.6}$ alloy in the amorphous state immediately after high pressure torsion does not experience martensitic transformations during cooling. After annealing, only B2↔R transformation occurs at grain sizes from 20 to 40 nm, while B2↔R↔B19' transformation occurs at the grain size interval from 40 to 80 nm and

Fig. 3.18 UFG structure of $Ti_{49.8}Ni_{50.2}$ alloy after eight passes of ECAP at 450 °C. Cross-section of the billet. (Reprinted by permission from Springer Nature [47], Copyright 2005)

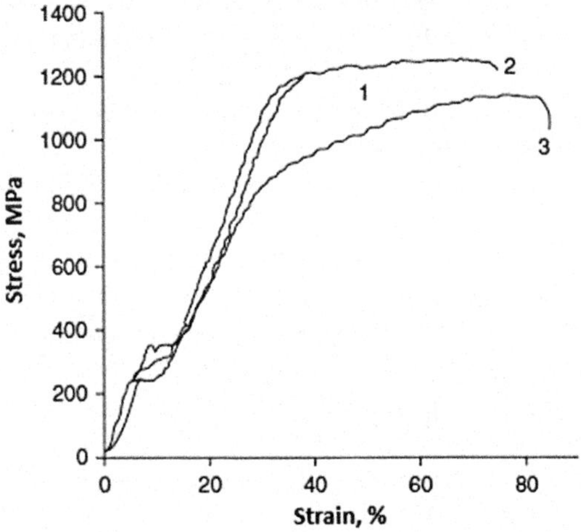

Fig. 3.19 Engineering tensile diagrams of $Ti_{49.8}Ni_{50.2}$ alloy after eight ECAP passes at 450 °C followed by 1 h annealing at 500 °C and 600 °C. (Reprinted by permission from Springer Nature [47], Copyright 2005)

there is B2↔B19' transformation in the coarse-grained state [45]. In this case, the symbol "R" denotes the intermediate martensitic phase with a rhombohedral unit cell, "B19'" is the martensitic phase with a monoclinic distorted orthorhombic (monoclinic) unit cell.

Processing of TiNi alloys by ECAP is usually conducted at elevated temperatures due to their difficulty of processing. For example, in $Ti_{49.8}Ni_{50.2}$ and $Ti_{49.4}Ni_{50.6}$ alloys, a typical microstructure formed by ECAP at 450 °C is characterized by the grain size ~250 nm (Fig. 3.18) [47]. The stability of this microstructure is retained up to the temperature of 500 °C.

The tensile strength σ_u of the $Ti_{49.8}Ni_{50.2}$ alloy subjected to ECAP after eight passes reached the value of 1250 MPa. At the same time, the yield strength reached the maximum value of 1150 MPa (Fig. 3.19) with the tensile strength being higher by 30% and the yield strength being twice as high as in the initial CG state. Additional annealing after ECAP at 500 °C for 1 h resulted in an increase in elongation-to-failure δ to ~50% and only a slight change in strength properties.

The formation of UFG structure results in the decrease of all characteristic temperatures (M_s, M_f, A_s and A_f) of martensitic transformations as compared to the quenched state [48]. The drop in the characteristic temperatures of martensitic transformations is about 20 ± 5 °C compared to the quenched state.

As discussed above, if the TiNi article is loaded with a load higher than the dislocation yield strength, the shape memory is not implemented. In this regard, the UFG structure, by increasing the yield strength, leads to an improvement in the two most important functional characteristics of shape memory alloys, namely, maximum reactive stress and maximum fully reversible strain. The maximum reactive stress obtained by ECAP, equal to 1100 MPa, is significantly higher than the values characteristic of both the coarse-grained state and the states after conventional thermomechanical treatments.

ECAP of $Ti_{49.4}Ni_{50.6}$ alloy leads to the occurrence of the superelasticity effect in a wide temperature range from 23 to 60 °C. Even at 20 °C, the value of superelasticity reaches ~7%, while the residual strain does not exceed 0.2%. In the CG $Ti_{49.4}Ni_{50.6}$ alloy, the effect of superelasticity at temperatures close to room temperature is not observed.

The application of titanium nickelide alloys with UFG structure opens up great prospects in medicine (see also Sect. 5.4) due to their strength properties that are considerably higher than those of the initial materials. The use of UFG alloys makes it possible to halve the cross-sections of the bearing elements of medical implants while keeping their reliability, significantly reduce the weight of the implants, which allows reducing the traumatism of operations, increasing the rate of tissue and bone recovery in the implant insertion area, and this, in its turn, reduces the treatment and recovery period of patients [49]. At the same time, TiNi alloys subjected to SPD have a unique combination of functional characteristics, which expands the prospects of their application.

The formation of UFG structures and their effect on the properties of the Cu–Al–Ni and Ni–Mg–Ga alloys, which also exhibit the shape memory effect, have been the object of active studies in recent years [50].

Suggested Reading

1. Valiev RZ, Islamgaliev RK, Alexandrov IV (2000) Bulk nanostructured materials from severe plastic deformation. Prog Mater Sci 45:103
2. Valiev RZ, Zehetbauer M, Estrin Y, Höppel HW, Ivanisenko Y, Hahn H, Wilde G, Roven HJ, Sauvage X, Langdon TG (2007) The innovation potential of bulk nanostructured materials. Adv Eng Mater 9:527
3. Sabirov I, Enikeev NA, Murashkin MY, Valiev RZ (2015) Bulk nanostructured materials with multifunctional properties. Springer, New York
4. Valiev RZ, Estrin Y, Horita Z, Langdon TG, Zehetbauer MJ, Zhu YT (2016) Fundamentals of superior properties in bulk nanoSPD materials. Mater Res Lett 4:1–21

Problems and Tasks

3.1. Physical Properties

1. List the microstructure parameters and explain their contribution to the achieving of the enhanced properties of UFG materials.

3.1.1 Electrical Conductivity

2. Within the framework of additive approximation, the contribution of various microstructure parameters to the electrical resistivity can be represented using the equation

$$\rho_R = \rho_i + \rho_p + \rho_d + \rho_v + \rho_{gb},$$

where ρ_i – the solid-solution component associated with scattering of conduction electrons on the atoms of alloying elements; ρ_p – the component due to the presence of particles of secondary phases, including Guinier-Preston zones, incoherent inclusions and coherent precipitations; ρ_d – dislocation component; ρ_v – vacancy component due to the presence of deformation vacancies; ρ_{gb} – grain boundary component of specific electrical resistivity.

Analyze the possible differences in the contributions of these components in the cases of similar CG and UFG metallic materials.

3. Suggest and justify the strategy of simultaneous achievement of high strength and electrical resistivity in electrotechnical metals and alloys.

3.1.2. Magnetic Properties

4. State and analyze the reasons of the influence of SPD on the magnetic properties of hard magnetic materials.
5. Analyze the role of grain boundaries in the engineering of magnetic properties of UFG soft magnetic metal alloys.

3.2. Chemical and Biomedical Properties of BNM Materials

3.2.1. Corrosion Properties

6. What are the microstructure parameters that can influence the corrosion resistance of metals and alloys? How their role may change in the case of UFG metallic materials?
7. Analyze the influence of the parameters of the microstructure formed at different stages of SPD on corrosion process.

3.2.2. Biomedical Properties

8. List the reasons for the high biocompatibility of pure titanium. Analyze the effect of grain size on biocompatibility.
9. Analyze the potential benefits of using UFG pure titanium to replace conventional titanium alloys in medicine.

References

1. Valiev RZ, Islamgaliev RK, Alexandrov IV (2000) Bulk nanostructured materials from severe plastic deformation. Prog Mater Sci 45:103
2. Valiev RZ, Zehetbauer M, Estrin Y, Höppel HW, Ivanisenko Y, Hahn H, Wilde G, Roven HJ, Sauvage X, Langdon TG (2007) The innovation potential of bulk nanostructured materials. Adv Eng Mater 9:527
3. Horita Z (2011) Production of mutifunctional materials using severe plastic deformation. In: Horita Z (ed) Proceedings of international symposium on Giant Straining Process for Advanced Materials (GSAM2010). Kyushu University Press, Fukuoka
4. Zehetbauer M, Groössinger R, Krenn H, Krystian M, Pippan R, Rogl P, Waitz T, Würschum R (2010) Bulk nanostructured functional materials by severe plastic deformation. Adv Eng Mater 12:692
5. Estrin Y, Zehetbauer MJ (2009) Niche applications of bulk nanostructured materials processed by severe plastic deformation. In: Zehetbauer MJ, Zhu YT (eds) Bulk nanostructured materials. WILEY-VCH Verlag GmbH & Co, Weinheim, p 635
6. Estrin Y, Vinogradov A (2013) Extreme grain refinement by severe plastic deformation: a wealth of challenging science. Acta Mater 61:782–817
7. Valiev RZ, Sabirov I, Zhilyaev AP, Langdon TG (2012) Bulk nanostructured metals for innovative applications. JOM 64:1134
8. Sabirov I, Enikeev NA, Murashkin MY, Valiev RZ (2015) Bulk nanostructured materials with multifunctional properties. Springer
9. Valiev RZ, Estrin Y, Horita Z, Langdon TG, Zehetbauer MJ, Zhu YT (2016) Fundamentals of superior properties in bulk nanoSPD materials. Mater Res Lett 4:1–21
10. Korznikov AV, Ivanisenko YV, Laptionok DV, Safarov IM, Pilyugin VP, Valiev RZ (1994) Influence of severe plastic deformation on structure and phase composition of carbon steel. Nanostruct Mater 4:159
11. Murashkin MY, Sabirov I, Sauvage X, Valiev RZ (2016) Nanostructured Al and Cu alloys with superior strength and electrical conductivity. J Mater Sci 51:33–49
12. Islamgaliev RK, Nesterov KM, Bourgon J, Champion Y, Valiev RZ (2014) Nanostructured Cu-Cr alloy with high strength and electrical conductivity. J Appl Phys 115:194301
13. Coey JMOD (2009) Magnetizm and magnetic materials. Cambridge University Press
14. Grössinger R, Sato R, Holzer D, Dahlgren M (2003) Properties, benefits, and application of nanocrystalline structures in magnetic materials. Adv Eng Mater 5:285–290
15. Stolyarov VV, Gunderov VD, Korznikova GF, Valiev RZ, Popov AG, Gaviko VS, Ermolenko AS (1997) Effect of severe plastic deformation on the structure and magnetic properties of Pr-Fe-B-Cu alloys. Phys Met Metallogr 83:173
16. Chbihi A, Savage X, Genevois C, Blavette D, Gunderov D, Popov AG (2010) Optimization of the magnetic properties of FePd alloys by severe plastic deformation. Adv Eng Mater 12:708–713
17. Straumal BB, Protasova SG, Mazilikin AA, Kogtenkova OA, Kurmanaeva L, Baretzky B, Schutz G, Korneva A, Zieba P (2013) SPD induced changes in microstructure and magnetic properties in the Cu-Co alloys. Mater Lett 98:217–221
18. Suehiro K, Nishimura S, Horita Z (2008) Change in magnetic property of Co-6.5Co alloy through processing by ECAP. Mater Trans 49:102–106
19. Scheriau S, Kriegisch M, Kleber S, Mehboob N, Grossinger R, Pippan R (2010) Magnetic characteristics of HPT deformed soft-magnetic materials. J Magn Magn Mater 322:2984–2988
20. Herzer G (1993) Nanocrystalline soft magnetic materials. Phys Scr T49:307–314
21. Ralston KD, Birbilis N (2010) Effect of grain size on corrosion: a review. Corrosion 66:075005
22. Miyamoto H, Yuasa M, Rifai M, Fujiwara H (2019) Corrosion behavior of severely deformed pure and single-phase materials. Mater Trans 60:1243–1255
23. Ralston KD, Birbilis N, Davies CHJ (2010) Revealing the relationship between grain size and corrosion rate of metals. Scr Mater 63:1201–1204

24. Noel JJ (2003) Effects of metallurgical variables on aqueous corrosion. In: Cramer SD, Covino BSJ (eds) ASM handbook 13A corrosion: fundamentals, testing, and protection. ASM International, Materials Park, p 258
25. Nazarov AA, Romanov AE, Valiev RZ (1996) Random disclination ensembles in ultrafine-grained materials produced by severe plastic deformation. Scr Mater 34:729–734
26. Lowe TC, Valiev RZ (2014) Bulk nanostructured metals in biomedical applications. In: Tiwari A, Nordin AN (eds) Advanced biomaterials and biodevices frontiers. Wiley-Scrivener, New York, pp 1–52
27. Valiev RZ, Prokofiev EA, Kazarinov NA, Raab GI, Minasov TB, Strasky J (2020) Developing nanostructured Ti alloys for innovative implantable medical devices. Materials 13:967–982
28. Williams DF (2003) On the mechanisms of biocompatibility. Biomaterials 29:2941–2953
29. Brunette DM, Tengvall P, Textor M, Thomsen P (2001) Titanium in medicine. Springer-Verlag, Berlin, p 1019
30. Lowe TC, Reiss RA, Illescas PE, Davis CF, Connick MC, Sena JA (2020) Effect of surface grain boundary density on preosteoblast proliferation on titanium. Mater Res Lett 8(6):239–246
31. Valiev RZ, Sabirov I, Zemtsova EG, Parfenov EV, Dluhoš L, Lowe TC (2018) Nanostructured commercially pure Ti for development of miniturized biomedical implants. In: Froes F, Qian M (eds) Titanium in medical and dental applications. Woodhead Publishing, Duxford, pp 393–418
32. Parfenov EV, Yerokhin A, Nevyantseva RR, Gorbatkov MJ, Liang CJ, Matthews CJL (2015) Towards smart electrolytic plasma technologies: an overview of methodological approaches to process modeling. Surf Coat Technol 269:2–22
33. Yerokhin AL, Parfenov EV, Matthews A (2016) In situ impedance spectroscopy of the plasma electrolytic oxidation process for deposition of Ca- and P-containing coatings on Ti. Surf Coat Technol 301:54–62
34. Rafieerad AR, Ashra MR, Mahmoodian R, Bushroa AR (2015) Surface characterization and corrosion behavior of calcium phosphate-base composite layer on titanium and its alloys via plasma electrolytic oxidation. Mater Sci Eng C Mater Biol Appl 57:397–413
35. Neoh KG, Hu XF, Zheng D, Kang ET (2012) Balancing osteoblast functions and bacterial adhesion on functionalized titanium surfaces. Biomaterials 33:2813–2822
36. Shin MH, Baek SM, Polyakov AV, Semenova IP, Valiev RZ, Hwang W, Hahn SK, Kim HS (2018) Molibden Disulfide surface modification of ultrafine-grained titanium for enhanced cellular growth and antibacterial effect. Sci Rep 8:9907
37. Venkatsurya PKC, Girase B, Misra RDK, Pesacreta TC, Somani MC, Karjalainen LP (2012) The interplay between osteoblast functions and the degree of nanoscale roughness induced by grain boundary grooving of nanograined materials. Mater Sci Eng C 32:330–340
38. Misra RDK, Nune C, Pesacreta TC, Somani MC, Karjalainen LP (2013) Understanding the impact of grain structure in austenitic stainless steel from a nanograined regime to a coarse-grained regime on osteoblast functions using a novel metal deformation-annealing sequence. Acta Biomater 9:6245–6258
39. Witte F (2010) The history of biodegradable magnesium implants: a review. Acta Biomater 6:1680
40. Witte F, Mantovani D (2011) Editorial. Mater Sci Eng B176:1595
41. Parfenov EV, Kulyasova OB, Mukaeva VR, Mingo B, Farrakhov RG, Cherneikin YV, Yerokhin A, Zheng YF, Valiev RZ (2020) Influence of ultra-fine grain structure on corrosion behaviour of biodegradable Mg-1Ca alloy. Corros Sci 163:108303
42. Pushin VG, Lotkov AI, Kolobov YuR, Valiev R Z, Dudarev EF, Kuranova NN, Dyupin AP, Gunderov DV, Bakach GP (2008) On the nature of anomalously high plasticity of high-strength titanium nickelide alloys with shape-memory effects: I. Initial structure and mechanical properties. Phys Met Metallogr 106:520
43. Pushin VG, Prokoshkin SD, Valiev RZ, Dudarev EF, Gunderov DV et al (2006) Alloys of nickel titanium with shape memory. In: Pushin VG (ed) Titanium nickelide alloys with shape memory. P I. Structure, phase transformations and properties. UrO RAS, Ekaterinburg, p 440

44. Buehler WJ, Gilfrich JV, Wiley RC (1963) Effect of low-temperature phase changes on the mechanical properties of alloys near composition TiNi. J Appl Phys 34:1475
45. Gunderov DV, Lukyanov AV, Prokofiev EA, Kilmametov AR, Pushin VG, Valiev RZ (2009) Mechanical properties and martensitic transformations in the nanocrystalline Ti49.4Ni50.6 alloy produced by high pressure torsion. Mater Sci Eng A 503:75–77
46. Huang JY, Zhu YT, Liao XZ, Valiev RZ (2004) Amorphization of TiNi induced by high-pressure torsion. Philos Mag Lett 84:183–190
47. Stolyarov VV, Prokofiev EA, Prokoshkin SD, Dobatkin SV, Trubitsyna IB, Khmelevskaya IY, Pushin VG, Valiev RZ (2005) Structural features, mechanical properties, and the shape-memory effect in TiNi alloys subjected to equal channel angular pressing. Phys Met Metallogr 100:608
48. Valiev R, Gunderov D, Prokofiev E, Pushin V, Zhu Y (2008) Nanostructuring of TiNi alloy by SPD processing for advanced properties. Mater Trans 49:97–101
49. Choynzonov EL, Gunther VE, Mukhamedov MR, Novikov VA et al (2013) Implants with shape memory in oncology. Tomsk State University, Tomsk
50. Lukyanov AV, Pushin VG, Kuranova NN, Svirid AE, Uksusnikov AN, Ustyugov YM, Gunderov DV (2018) The influence of thermal and mechanical treatment on structural-phase transformations in Cu-14Al-3Ni alloy with a shape memory effect subjected to high-pressure torsion. Phys Met Metallogr 119:393–401

Chapter 4
Synthesis of Bulk Nanostructured Metal Systems

The earlier chapters describe the principles of SPD techniques, fundamental knowledge required for nanostructuring, and general UFG microstructures, and the enhanced mechanical properties and additional functionalities of UFG materials processed by SPD. While significant efforts were placed for understanding the produced nanocrystalline microstructures and their advanced properties in the last two decades, a new research focus was developed for the utilization of conventional SPD techniques, with some modifications in procedures to synthesize advanced, hybrid nanocrystalline metal systems. Here, hybrid nanocrystalline materials refer to composites consisting of two or more constituents at the nanometer level or the nano-to-micro level leading to a gradient or heterogeneous microstructure formation. Accordingly, this chapter summarizes the recent developments in strategies for the formation of bulk nanostructured metal systems and UFG materials from powders and dissimilar bulk metals. The further sections discuss the uniquely tuned nanostructures, heterostructures, and multilayered laminates and their improved mechanical properties and additional functionalities of the hybrid nanocrystalline materials after SPD processing. Recent developments in architecturing of heterostructured nanostructures and their possible future applications are also discussed by considering SPD techniques and nanostructures synergy.

4.1 Strategies for the Formation of Bulk Metal Systems

4.1.1 Powder Consolidation by SPD

The synthesis of bulk materials from powders has attracted significant attention, while it brings several things that conventionally manufactured solid materials, e.g., casting, can avoid. In practice, powder metallurgy often has a difficulty to introduce bulk materials with full density. In addition, powder consolidation in conventional

R. Z. Valiev et al., *Ultrafine-Grained Materials*, The Minerals, Metals & Materials Series, https://doi.org/10.1007/978-3-031-31729-3_4

powder metallurgy has been operated using sintering at high temperatures, which is not economical and leads to a loss of the finer microstructure of the initial powders after the consolidation. Therefore, it is reasonable to develop an alternative approach for powder consolidation in order to achieve a high density of bulk materials at lower processing temperatures.

Mechanical consolidation of powders and particles into bulk ultrafine and nanostructured metal systems has been developed in the last three decades by utilizing the SPD techniques [1]. The advantages of this approach involve a high degree of shearing between the powders during consolidation by SPD, so that the resultant bulk metals involve not only successful bonding of powders achieving high bulk density but also unique phase transformations of phases, such as the dissolution of minority phases as well as the formation of metastable phases. Consolidation of powders by SPD was first published in 1991, where oxidized iron powders, including a total volume fraction of 18% FeO and Fe_3O_4, were consolidated under pressure and torsion straining [2]. These two iron oxide phases are gradually dissolved into the Fe matrix during intensive cold working of the torsional deformation under pressure. In this experiment, additional post-processing annealing at 600 °C for 40 min led to the dissolution of all Fe_3O_4 into FeO and Fe phases.

The powder consolidation by SPD, especially HPT, was proven as the most effective procedure in comparison with conventional uniaxial powder compaction for producing a bulk nanostructured Cu–50 wt.% Ag solid solution alloy [3]. In practice, torsion straining under 3 GPa compressive pressure introduced an Ag–Cu solid solution alloy having an *f.c.c.* phase irrespective of the Ag–Cu powder preparations by mechanical alloying or inert gas condensation, leading to a pre-alloyed powder, while the initial powders prepared by these two powder preparation techniques have two separate α-Cu and α-Ag phases. Thereafter, ball-milled metal powders, consolidated by the severe plastic torsion straining (SPTS) procedure [4, 5] that is an earlier terminology of HPT, have been studied actively for the synthesis of bulk nanostructured hybrid metals. In practice, these series of reports demonstrated the synthesis of bulk nanocomposites from the Cu–SiO_2 and Al–Al_2O_3 powder mixtures by the SPTS powder consolidation at over 6 GPa and the three produced disk samples are shown in Fig. 4.1 [6], where a one-cent US coin is placed lower for

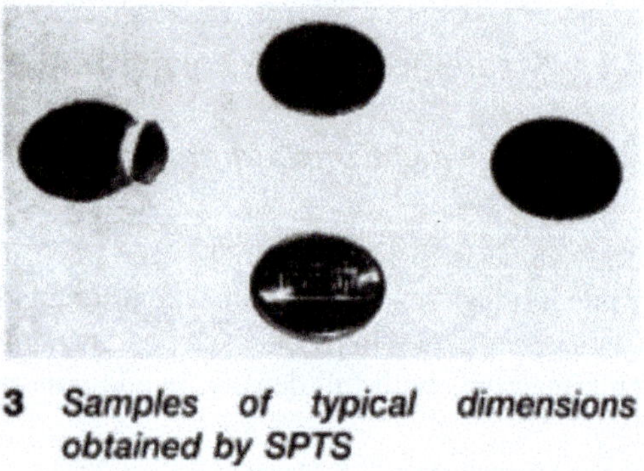

Fig. 4.1 A photo of produced disk samples after powder compaction. (Reprinted by permission from Springer Nature [6], Copyright 1998)

comparison purposes and one of the processed disks has broken into two parts due to its brittleness and a possibility of porosities initiating macro-cracks.

The powder consolidation by HPT has been often conducted by using one selected metal and alloy element as a ball-milled nanocrystalline powder, such as Al–Fe [7] and Al–Mg alloys [8], Co [9], Ni [10], and Ti [11]. Other studies used machine chips of Cu [12] and an Al–8Si–3Cu alloy [13], gas atomized amorphous powders of $Al_{90}Fe_5Nd_5$ [14] and $Al_{90}Fe_7Z_3$ [15], and several oxide powders [16], including α-Al_2O_3 [17], $BaTiO_3$ (a group of perovskite oxides) [18], SiO_2, and VO_2 [19]. Successful formation of a bulk disk sample from machine chips is shown in Fig. 4.2 for an $AlSi_8Cu_3$ alloy, where (a) machine chips were (b) cold compacted, and (c) processed by HPT under 8 GPa for ten turns to have a disk sample [13]. Moreover, as was shown in Fig. 4.1 for the Cu–SiO_2 and Al–Al_2O_3 powder [6], HPT can be applied as mechanical mixing during the consolidation for producing metal-ceramic nanocomposites and hybrid systems from a combination of metal powders and ceramic powders, such as Ti and TiO_2 powders [11], Al powders with carbon nanotubes [20], Al powders with fullerenes (a mixture of C_{60} and C_{70}) [21], Co and NiO powders [22], Co and α-Al_2O_3 powders [23], and $Al_{84}Y_3Ni_8Co_4Zr_1$ and SiC powders [15].

In addition to HPT processing, several SPD techniques have been utilized for powder consolidation toward the introduction of bulk nanostructured materials. For example, ECAP has been applied to consolidate pre-cold isostatic pressed powders of Al_2O_3–3mass%Fe–5mass%Ni [24], and gas atomized powders of Al-based [25], Cu-based [26], Cu powder with carbon nanotubes [27], and Zr-based amorphous metals [28, 29] at over 400 °C for the introduction of an amorphous metal-matrix nanocomposite (AMMC). In order to produce a high-pressure atmosphere for better powder bonding under uniform shear by ECAP into a large sample volume, back pressure equal-channel angular consolidation (BP-ECAC), as shown in Fig. 4.3a,

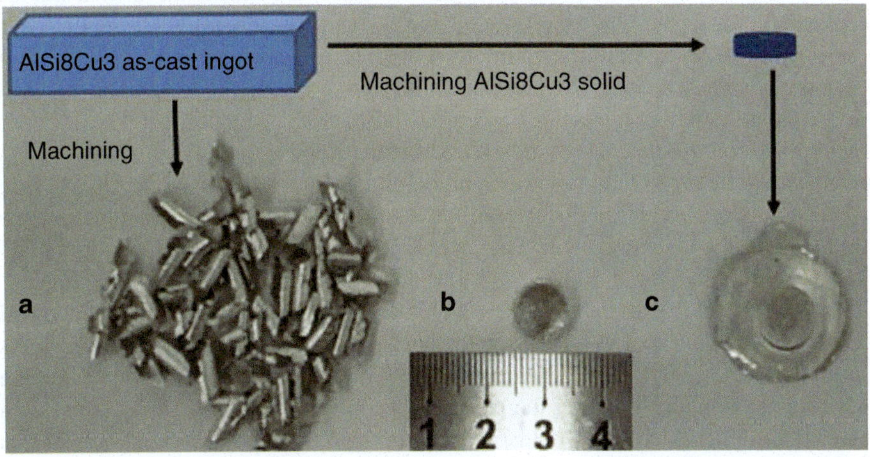

Fig. 4.2 Macrographs of the (**a**) as-machined chip, (**b**) cold compacted chip sample, and (**c**) HPT-processed AlSi8Cu3 alloy samples. (Reprinted from [13], Copyright 2013, with permission from Elsevier)

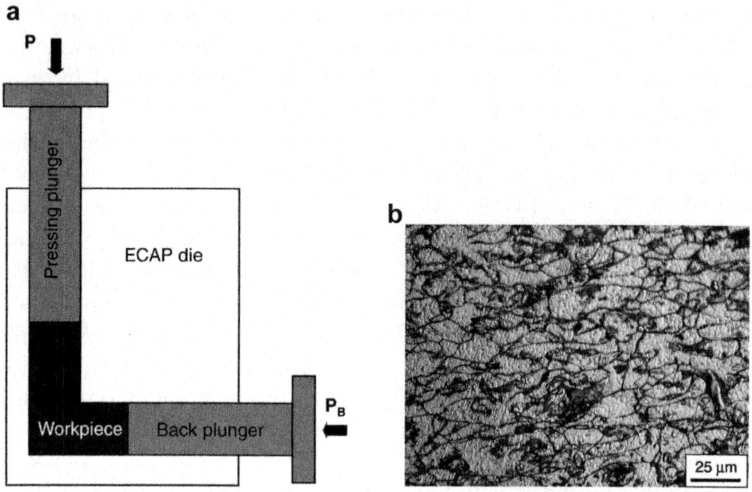

Fig. 4.3 Schematic drawings of (**a**) BP-ECAC, (**b**) the optical microscope photo of the Al powders consolidated by BP-ECAC. (Reprinted from [30], Copyright 2005, with permission from Elsevier)

was introduced for enabling the consolidation of commercial-purity Al powders at a temperature of 100 °C for 1 pass under a back pressure of 50 MPa [30]. Good bonding with grain refinement is observed without any apparent pores between the consolidated particles, as shown in Fig. 4.3b [30], and the refined average grain size of 7.2 μm was considerably finer than the initial particle sizes of 60–100 μm.

Moreover, torsion extrusion (TE) for powder consolidation is also feasible to introduce both reasonable high pressure under torsion as well as extrusion components and the atomized commercial-purity Al powder was solidified by this process at 350 °C [31]. Schematic drawings of the TE facility are shown in Fig. 4.4a and the optical micrographs showing representative porosities at the different sample zones during the TE process are shown in Fig. 4.4b [31]. It is apparent in the micrographs that the powders after the TE process demonstrated nearly fully dense microstructure, while the report showed the consistent TE facility without the torsion component failed to introduce a fully-dense microstructure.

It should be noted that most metal particles are covered by an oxide layer having several nanometers to micron thicknesses on the material surfaces that prevent the bonding of powders in general. Thus, in conventional powder metallurgy, the powder consolidation for introducing bulk metals requires high processing temperatures, usually higher than 0.5 T_m, to accelerate the diffusion of both the main metal elements and their oxides. However, the conventional procedure often leaves pores in the condensed bulk materials. In order to reduce the porosity for increasing the density of the bulk solid after the powder consolidation, sintering processing often involves high pressure, cold/hot isostatic pressing (CIP and HIP) [32, 33] and/or an electric field in a field-assisted sintering technique (FAST) [34].

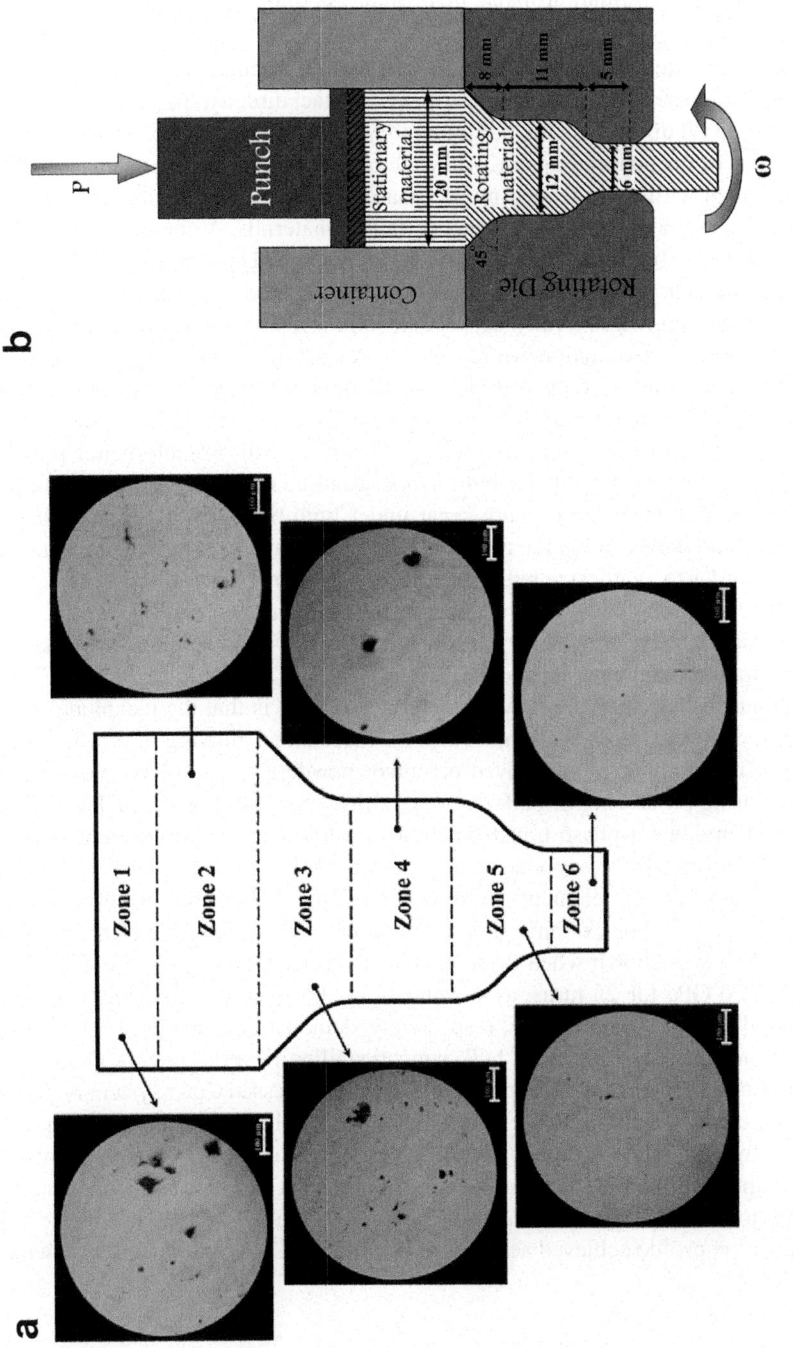

Fig. 4.4 Schematic drawings of **(a)** TE facility and **(b)** the optical micrographs showing representative porosities at the different sample zones during the TE process. (Reprinted from [31], Copyright 2010, with permission from Elsevier)

By contrast, the mechanism for powder consolidation through SPD involves dynamic plastic deformation under hydrostatic pressure and a simple but severe shear on powders and particles with their oxide layers. In practice, the oxide layers are hard and brittle and are fragmented and fractured during severe shear deformation [35]. Thus, the cores of the powder metals contact directly, they bond, they flow reasonably to fill up the existing pores and they deform during the refinement of the domain size of the material matrix, while the broken pieces of the fine metal oxides are distributed within the refined matrix phase. This procedure ultimately leads to high density in the resultant bulk nanocrystalline materials. A comparison of powder consolidation between conventional sintering and SPD processing is elsewhere [36], where a bulk solid, including porosities, can be produced after hours of conventional sintering, while a fully-dense bulk solid is available instantaneously after the SPD powder consolidation. Moreover, even under SPD processing, an introduction of high friction between particles can be possible when the initial materials have large powder sizes, irregular particle shapes, and low hardness of powders. In addition, high processing temperatures and pressures can promote better powder consolidation toward fully dense bulk nanocrystalline materials. Thus, due to the principle of HPT leading to severe shear under high pressure, it is reasonable to recognize that HPT is one of the most efficient SPD techniques for powder consolidations in order to synthesize bulk nanocrystalline hybrid materials. The effects of the particle characteristics (size, shape, strength, and surface conditions) and processing conditions on the powder consolidation by SPD (temperature and pressure) are the important factors to consider in the procedures [36].

A uniqueness of applying powder materials to HPT is that the technique is not only for the powder consolidation but also for mechanical alloying of powders with dissimilar metal elements or alloyed precursor powders by the severe mechanical mixture during processing of bulk nanocrystalline solids. In practice, HPT powder processing initiates a phase transformation to introduce non-equilibrium phases, including intermetallic compounds and supersaturated solid solutions [35, 37], the segregation of alloying elements in crystalline defects [38], and the formation of amorphous phases. For example, atomic diffusion of an immiscible elemental pair of Cu and Cr was shown when these powders were consolidated by HPT under a pressure of 6 GPa for 25 turns, as shown in Fig. 4.5 [37]. Figure 4.5a, b of TEM bright field and dark field images, respectively, demonstrates that the HPT powder compaction introduced a $CuCr_{43}$ bulk nanocrystalline material having an average grain size of 10–20 nm. A 3D mapping of the HPT-processed CuCr system is shown in Fig. 4.5c with an atomic scale in a volume of $12 \times 12 \times 50$ nm^3 by atom probe tomography. The three-dimensional mapping proves a significant amount of Cu atoms in the Cr grains, a Cu grain is necked (marked A) and further forms a nanoscale tip (marked B). The dissolution of Cu atoms into a Cr grain was confirmed by a composition profile achieved across a representative Cu/Cr interface, as shown in Fig. 4.5d [37].

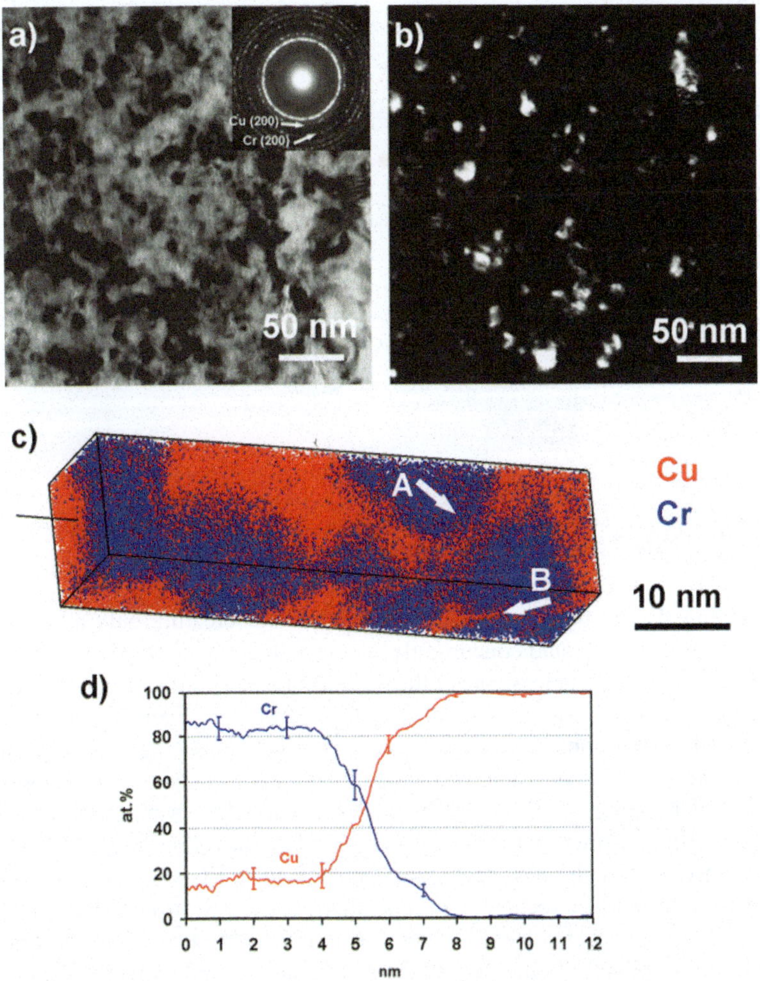

Fig. 4.5 (**a**) TEM bright field image and related SAED pattern and (**b**) dark field image of a Cu-Cr powder mixture forming a $CuCr_{43}$ bulk nanocrystalline material processed by HPT through 25 turns, and (**c**) 3D reconstruction of the $CuCr_{43}$ nanocomposites in a volume of $12 \times 12 \times 50$ nm^3, where the necking of a Cu grain (a) and the resulting nanoscale tip (**b**) are shown with arrows. And (**d**) composition profile computed across a Cu/Cr interface. (Reprinted from [37], Copyright 2008, with permission from Elsevier)

4.1.2 Bulk-State Reactions by SPD

As was shown in the previous section, powder consolidations by SPD is an attractive solid-state-processing approach for producing bulk nanocrystalline metals and hybrid materials. However, these methods require sufficient amounts of powders to produce the bulk systems to achieve the final consolidated products. Considering the solid-state joining of dissimilar metals, both compound casting [39] and friction

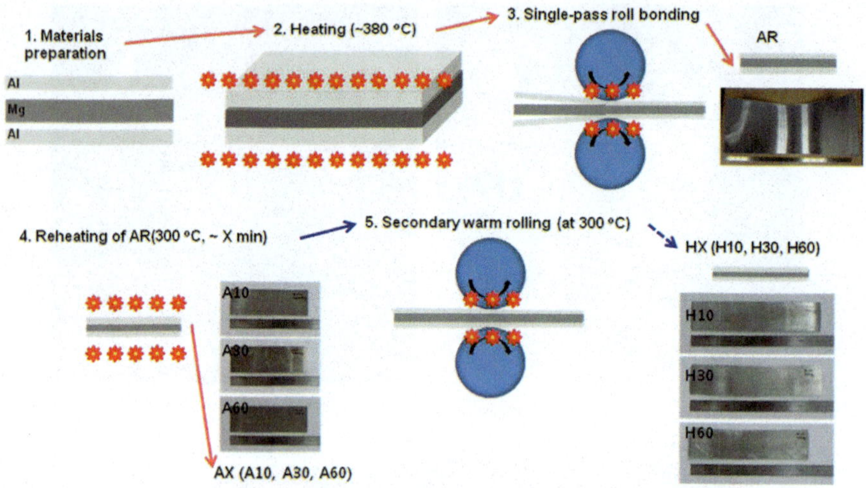

Fig. 4.6 A schematic drawing of processing steps of warm rolling for three-ply Al/Mg/Al clad sheets. (Reprinted from [49], Copyright 2014, with permission from Elsevier)

stir welding (FSW) [40] are possible approaches for creating bulk alloy systems from dissimilar metals and concurrently acquiring fine crystalline structures, but in practice, these require at least one of the metals to be in a liquid state for casting. For a direct bulk synthesis of nanocrystalline alloy systems, mechanical bonding of dissimilar bulk metals under severe plastic deformation is an alternative approach from simple and inexpensive solid precursors [41]. Representative presentations of mechanical bonding of dissimilar metals in a bulk-state were demonstrated by solid-state roll-bonding methods of general cold/warm rolling [42], which can be further improved by the SPD techniques of ARB [43], ECAP [44], HPT [45, 46], and a further developed technique of tube high-pressure shearing (t-HPS) [47].

Processing by roll cladding [48] and warm rolling [49] involves a single or more steps of conventional thermomechanical rolling at elevated temperatures and produces a multilayered plate from dissimilar metals. The resultant plate shows the microstructure with micron-scale grains. When the processes involve rolling at elevated temperatures or heat treatments between the rolling steps, the interfaces of the dissimilar metal plates form several layers of one or more intermetallic compound phases due to diffusion of the elements, leading to metal-matrix composite materials exhibiting high strength where these layers are not uniform. Figure 4.6 shows the detailed processing steps of two-step roll bonding of Al/Mg/Al clad sheets [49]. In this procedure, before and after the second rolling step at 300 °C, the Al/Mg/Al plates were annealed for different time periods of 10, 30, and 60 min. The reasons for the additional heating processes are due to strengthening of the interfacial bonding of the dissimilar metals by diffusion.

In order to further refine the grain size and improve the mechanical properties of the hybrid sheets produced by roll-bonding, the ARB process has been utilized for the introduction of multilayered laminates by a severe bulk-state reaction. In the

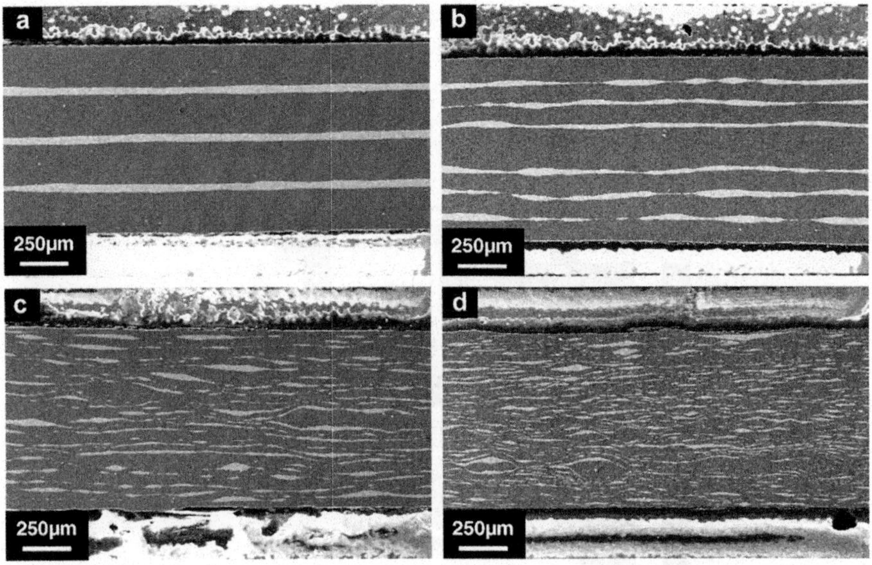

Fig. 4.7 Microstructural development of multi-layered Al (dark layers)/Cu (blight layers) composite through (**a**) the initial bonding of Al and Cu by degreasing of bonding surfaces and stacking, (**b**) first cycle, (**c**) second cycle, and (**d**) fifth cycle by ARB. (Reprinted from [43], Copyright 2008, with permission from Elsevier)

ARB process, two dissimilar sheet materials are stacked, roll-bonded at ambient to elevated temperatures, and then the processed sheet is cut in half, the surfaces are cleaned, two bonded sheets are stacked together, and the process is repeated. Figure 4.7 shows the microstructural development of a multilayered Al (dark layers)/Cu (bright layers) composite through (a) the initial bonding of Al and Cu by degreasing of bonding surfaces and stacking, (b) first cycle, (c) second cycle and (d) fifth cycle by ARB at room temperature [43]. As is apparent from Fig. 4.7a, good bonding of dissimilar metals by conventional rolling and ARB is possible only when the oxide films and contaminations of each sheet are removed by brushing and other special treatments on the surfaces prior to roll-bonding. Such surface preparations before bonding provide direct contact of the dissimilar metal elements.

The phase refinement and fragmentation of Cu phases are apparent with increasing ARB cycles in Fig. 4.7, but the observed grain sizes are often at the sub-micron level. Several investigations in ARB processing of dissimilar metals provided the knowledge that a successful solid-state reaction of Al and Cu plates and other metal combinations toward a better mixture of the phases with increasing ARB cycles requires a special sample setup of a hard phase having thinner plates between soft materials having thicker plates. Moreover, characterization of the fine-grained metal plates after ARB generally reveals anisotropic plastic behavior, which depends on the rolling direction, and which varies in the through-thickness direction as well [50]. Such anisotropic plastic flow also occurs in the multilayered roll-bonded laminates [51].

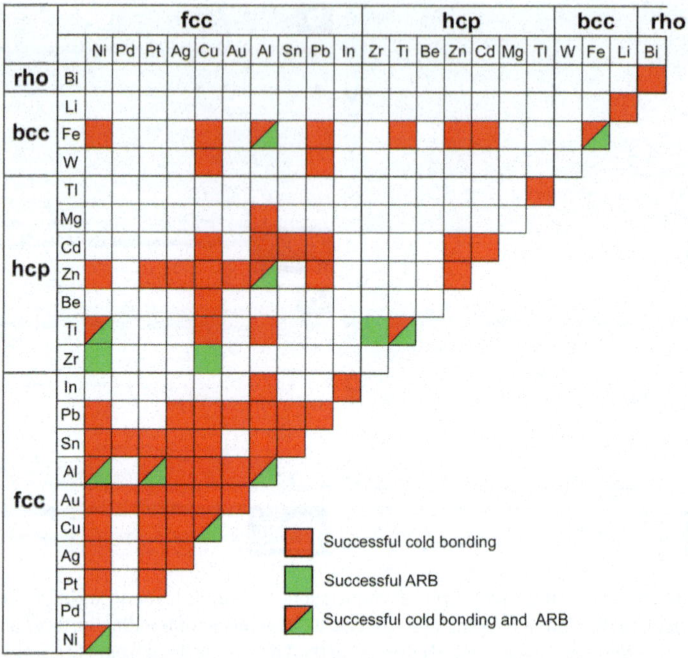

Fig. 4.8 Map showing a series of metal combinations for successful cold bonding by rolling and/or ARB. (Reproduced with permission [52])

Fig. 4.9 Optical micrographs of Cu–Nb ARB material from initial bonding step (far left) through seven bonding steps (far right). (Reprinted from [53], Copyright 2015, with permission from Elsevier)

A bonding map, as shown in Fig. 4.8 [52], describes examples of metal pairing where the color-marked pairs are successfully bonded through clad-bonding by cold rolling and/or ARB. The map should be updated, but this type of map provides an indication that *f.c.c.* metals are often chosen for the bulk-state bonding of dissimilar metals due to their reasonable initial ductility, and thus formability, rather than metals having other lattice structures during plastic deformation.

An example of the formation of the multilayered nano-laminates was shown by ARB processing by applying an immiscible metal combination of Cu having an *f.c.c.* structure and Nb having a *b.c.c.* structure. The cross-sections of the processed plates involving different thicknesses of initial plates are shown in Fig. 4.9 [53],

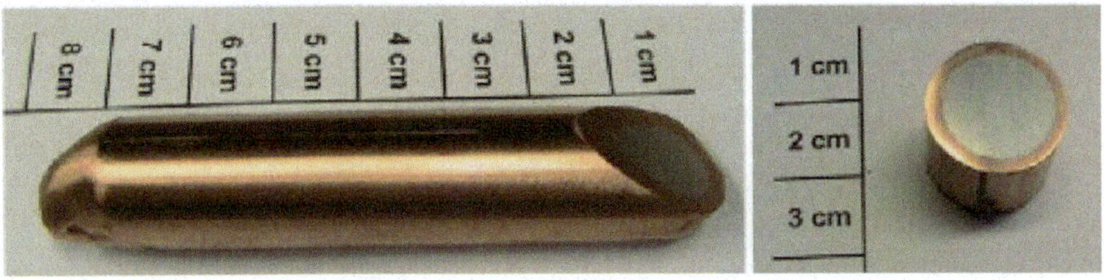

Fig. 4.10 An Al and Cu bimetal rod processed by ECAP for one pass at (left) side and (right) a cross-section. (Reprinted from [44], Copyright 2007, with permission from Elsevier)

where initial thicknesses of Cu (yellow) and Nb (gray) plates are equivalent within each plate. It is apparent that the immiscible elements maintain reasonably flat interfaces without necking of the phases under ARB. Decreasing thicknesses of these metals from a few millimeters to less than 10 nm led to an increase in an interfacial areal fraction leading to an exceptional hardness of ~4.5 GPa and yield strength of ~2 GPa, and ultimately processing by ARB introduced Cu–Nb nano-laminates without any metastable formation in a bulk solid.

Processing by ECAP has been used for the formation of bimetallic rods or billets through mechanical bonding. The first example was shown for bonding of solid Al and Cu and the billet after 1 ECAP pass at 350 °C, as shown in Fig. 4.10 [44]. In practice, a small cylindrical Al billet having a diameter of 10 mm and a length of 70 mm covered by an outer layer of Cu having inner and outer diameters of 10.1 mm and 13.9 mm, respectively, are processed by ECAP with a 90° die at 350 °C for up to 2 passes. Very good bonding at the interface of Al and Cu is visible. Not only increasing the shear strength of the Al–Cu bimetal with increasing numbers of ECAP passes but also the maximum shear strength of the bimetal processed by ECAP is 12.5% higher than after general extrusion imposing a consistent equivalent strain with ECAP.

A combination of Al/Cu bonding by ECAP was studied to further understand the crack formation due to the shearing of Al and Cu phases, the influence of the Cu outer layer thickness on the homogeneity of the microstructure, and the population of high-angle grain boundaries when processing at room temperature [54] as well as the intermetallic formation at the interfaces when ECAP processing was conducted at 200 °C [55]. Moreover, Cu/Ni [55], Cu/Nb [56], and Al/Steel bimetallic rods are reported for bonding of the bulk metals and refining the microstructure without segregations at the interfaces, leading to an improvement in mechanical properties and functionalities, including electrical conductivity [57]. During processing by ECAP, these materials experience not only the general phenomena of simple shear and hydrostatic pressure but also the process on bimetallic materials can create frictional shear stresses at the interfacial region which affect the microstructure and thus the properties of the ECAP-induced hybrid nanocrystalline alloys.

Accordingly, modifying the design of dissimilar metals can change the hardness distribution of the resultant bimetal billet after ECAP. In practice, lowering the wall thicknesses of the dissimilar phases through architecturing the bimetal formation can introduce a homogeneous distribution of frictional shearing within the bulk sample of the two dissimilar elements of Cu and Nb [56]. Specifically, Cu/Nb

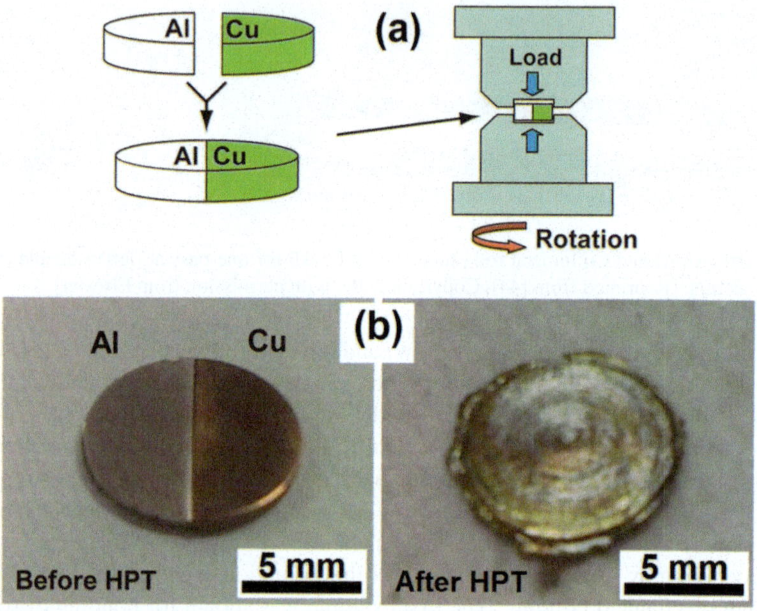

Fig. 4.11 (**a**) Schematic illustration of solid-state reaction of Al and Cu semicircle samples by HPT and (**b**) appearance of Al-Cu sample before and after HPT for 100 turns. (Reprinted from [45], Copyright 2013, with permission from Elsevier)

bimetal rods having ~12 mm diameter were prepared with an Nb core surrounded by two Cu tubes and Nb tubes in the order of Cu/Nb/Cu and 85 segments of Nb filament surrounded by Cu and Nb tubes that are bundled in a Cu matrix. The associated hardness values at several local locations are in the plots below the cross-sectional photos of the Cu/Nb samples after ECAP for an equivalent strain of 2.54. There is a hardness increase after ECAP in the copper matrix, whereas the hardness of the Nb remained reasonably unchanged in both designs. However, the design with a thinner Cu phase increases the hardness of Cu more than the Cu with thick wall thickness, and the thin Cu wall leads to fewer hardness fluctuations across the diameter of the bimetal Cu/Nb rod.

The bulk-state reaction of bulk dissimilar metals can be introduced by HPT by applying two or more different metals and alloys in different sample shapes. The first experiment was reported using a full circular disk consisting of Al and Cu semicircular disks, as shown schematically in Fig. 4.11a, and the disk was processed by HPT under a compressive pressure of 6.0 GPa for 100 turns at room temperature, as shown in Fig. 4.11b [45]. The processed disk has a mixture of Al- and Cu-rich ultrafine grains having 50–500 nm grain sizes as well as nano-scale intermetallic compounds of Al_2Cu and Al_4Cu_9. As was shown in an earlier section for powder consolidation, the bulk-state reaction of dissimilar bulk metals by HPT forms intermetallic phases, while the mechanism of the intermetallic phase nucleation is due to the accelerated diffusion under severe torsion and high pressure aided by the presence of vacancies, dislocations, and grain boundaries.

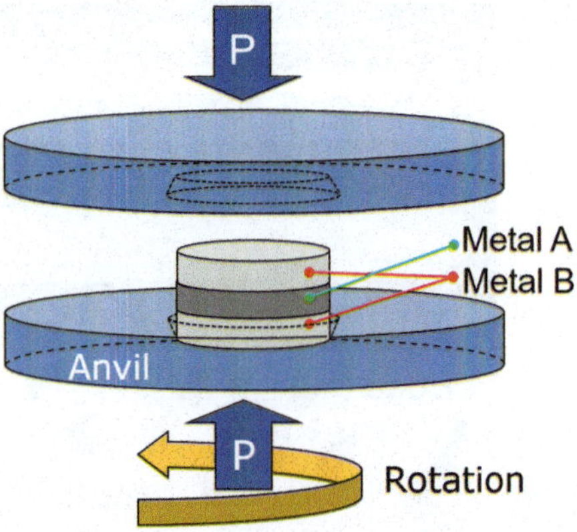

Fig. 4.12 Schematic drawing of the bulk-state materials bonding by HPT. (Reprinted from [46], Copyright 2015, with permission from Elsevier)

A similar study of the bulk-state reaction by HPT was shown by applying a whole disk made of four quarter-circle disks of Al and Cu alloys [58]. Moreover, the HPT-induced bulk-state reaction became common by utilizing easier sample preparations by directly stacking a few circular disk samples on the HPT anvils without any surface brushing treatment between the samples, as shown schematically in Fig. 4.12 [46]. In practice, a series of experiments was conducted by the bulk-state bonding by HPT of separate Al and Mg disks, for example, an Al-1050 aluminum alloy and a ZK-60 magnesium alloy in Fig. 4.12, by stacking them alternately in the order of Al/Mg/Al and processed under 6.0 GPa at room temperature. A combination of Al and Mg formed two intermetallic compounds of Al_3Mg_2 and $Al_{12}Mg_{17}$ within the nanocrystalline Al-rich phase during grain refinement by the HPT process at room temperature for 20 turns, thereby demonstrating the introduction of a metal-matrix nanocomposite. Ultimately, all Mg dissolved into the Al matrix, leading to a bulk metastable Al solid solution with a supersaturated solid solution state after 100 turns [59]. Such high numbers of HPT turns create ultra-severe plastic deformation [60], which helps to dissolve minority phases to the major and matrix phase in the ambient temperature processing. After 100 HPT turns on the Al/Mg/Al disks, an exceptional supersaturation of Mg solutes with a content of ~38.5 at.% was observed without any second-phase or intermetallic phases within the metastable Al solid solution. A comparison of cross-sections of the Al/Mg/Al disks after HPT for 1 turn and 100 turns under 6 GPa is shown in the upper section of Fig. 4.13 [59]. Exceptional grain refinement was achieved homogeneously throughout the metastable alloy having nano-scale grain sizes of 35–40 nm, as shown in the lower section of Fig. 4.13. This approach of stacking dissimilar materials for producing nanocrystalline hybrid materials by HPT has been studied with a variety of material combinations [61], such as Al/Cu, Al/Fe, Al/Ni, Al/Mg, Al/Ti, Al/Mg/Cu/Fe/Ti, Ag/Cu, Ag/Ni, Cu/Ta, Cu/ZnO, Fe/V, V–10Ti–5Cr/Zr–2.5Nb, Zn/Mg, and Nb/Zr.

It should be noted that the immiscible metal combinations of the Ag/Ni and Nb/Zr systems were mechanically mixed by HPT with a consistent sample setup with the model shown in Fig. 4.12 under 5 GPa for up to 100 turns [62]. While the Ag/Ni alloy system remained immiscible after 100 HPT turns, the Nb/Zr alloy introduced the diffusion of these elements and ultimately transformed into a supersaturated

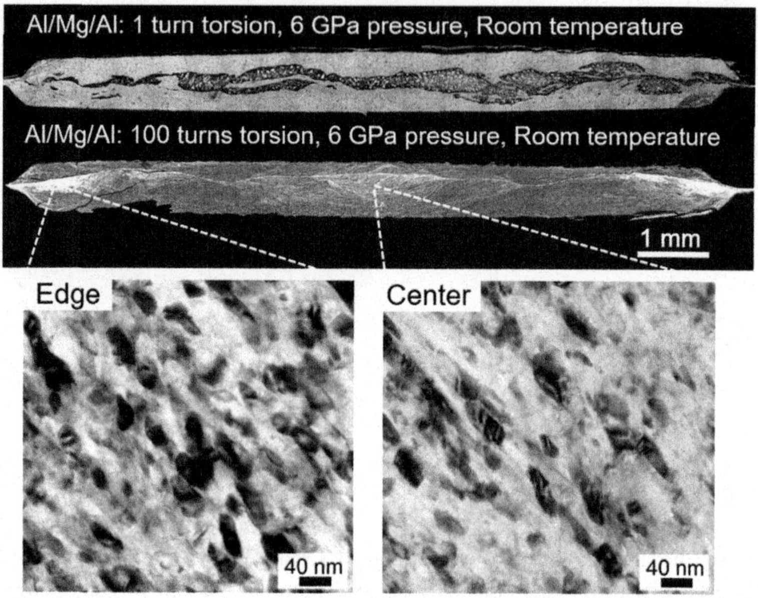

Fig. 4.13 (Upper) Cross-section of the Al/Mg/Al disk after HPT for 1 turn and 100 turns under 6 GPa. (lower) TEM micrographs taken at the disk edge and center proving microstructural homogeneity throughout the disk material. (Reprinted from [59], Copyright 2019, with permission from Elsevier)

solid solution having a single *b.c.c.* structure at the measured disk surface. The gradual change toward supersaturation of Nb and Zr through 100 HPT turns can be seen in X-ray diffraction patterns of the Nb–Zr alloy system after HPT for 2, 10, 20, and 100 turns [62]. Thus, it is apparent that HPT processing is feasible to provide an extreme mechanical alloying effect during nanostructuring even in the immiscible metal systems.

In contrast to the powder metallurgy approach, the bulk-state reaction is initiated at the flat interfaces between the dissimilar bulk metals. The formation of metastable phases, including intermetallic phases and supersaturated solid solution phases, occurs in the vicinity of the interfaces under severe shear by HPT. Thus, the deformation mechanism in the macro-scale of the stacked dissimilar metals during HPT can be expressed by the formation of tribomaterials [63], where it requires the severe shear and sliding of dissimilar metals at the interfaces.

In the nano-scale mechanism of the bulk-state reaction, the microstructural flow is caused by the plastic instabilities attributed to local blocking of simple shear flow induced by HPT, and Fig. 4.14 shows the unique plastic deformation of the stacked metal alloys under HPT in (a) and (b) schematic and (c) and (d) modeling [64]. Specifically, Fig. 4.14a, b show the schematic illustrations of the sample setup and the shear flow, respectively, in a stacked metal system when a hard metal is placed between the soft metal solids under simple shear, where in (a) the yellow arrows show the anvil motions leading to pure shearing in the early stage of deformation,

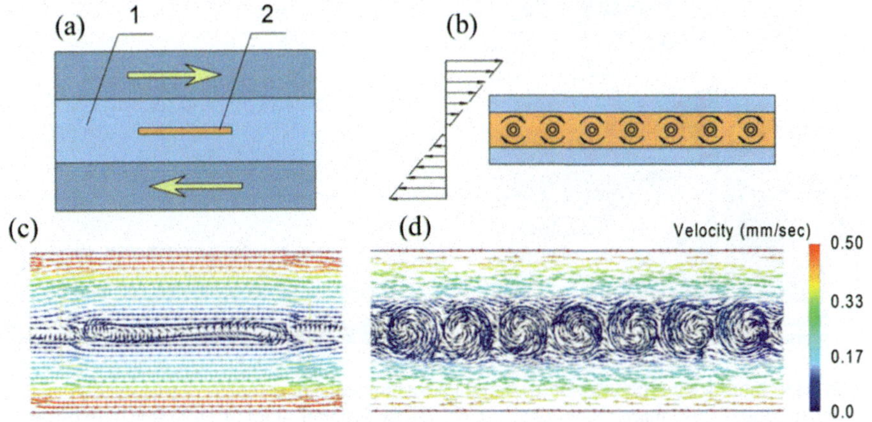

Fig. 4.14 A model of simple shear with local blocking of shear deformation: (**a**) The yellow arrows show the anvil motions, and the medium "1" with a layer of a hard material is located in the middle at a position "2" is subjected to simple shear in the horizontal direction, (**b**) velocity profile of the medium outside of a stronger layer and representation of the layer that contains the velocity discontinuity in the form of a chain of vortices where the vortex axes are perpendicular to the plane of the figure, and (**c**) and (**d**) demonstrate the metal flow model which is equivalent to (**a**) and (**b**), respectively, with the color-coded flow velocity. (Reprinted from [64], Copyright 2018, with permission from Elsevier)

and it subsequently causes a series of simple shears in (b). Figure 4.14c, d demonstrate the metal flow model which is equivalent to (a) and (b), respectively, with the color-coded flow velocity. The model in Fig. 4.14c shows that the severe shear by HPT can be terminated at the interfaces of the dissimilar metals since the yield strengths of the materials are different, leading to a velocity gradient across the material thickness. Subsequently, the mechanically bonded alloy system shows a velocity discontinuity in the form of a chain of vortices as shown in Fig. 4.14b, d. Thus, the bulk-state reaction by bonding dissimilar materials under HPT introduces large volumes of nanocrystalline tribomaterials during mechanical mixing and grain refinement through plastic instability.

Although there are many varying procedures which can enable the initiation of a solid-state reaction, here we give one more example of the bulk-state reaction. The formation of multilayered laminates in a tube sample was shown by tube high-pressure shearing (t-HPS) [65]. Processing by t-HPS uses a sample in the form of a tube, where the outer surface of the sample is surrounded by an outer cylinder, while the inner surface is fixed by a central mandrel, the sample is fixed by a set of pressure rings to produce high hydrostatic pressure in a tube height direction, and it is sheared by rotating the outer cylinder with respect to the inner surface. Using this processing technique, two different elements of Pb and Sn in the form of quarter tubes were placed alternately to construct one complete tube and these were processed by t-HPS under 2 GPa at room temperature for up to 8 turns. Figure 4.15 shows (a) an initial Pb (dark)/Sn (bright) bimetallic tube sample consisting of four parts of quarter tubes, (b) after 1/4 turn, (c) after 1/2 turn, and (d) after 8 turns by

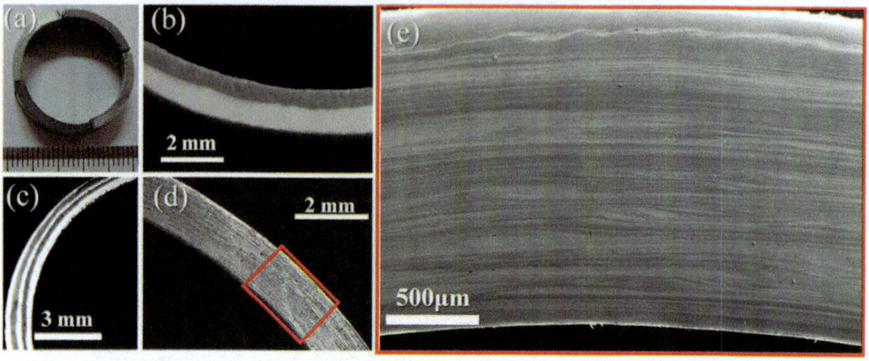

Fig. 4.15 Interface morphology of (**a**) initial Pb/Sn bimetallic tube sample consisting of four parts of quarter tubes (two quarter tubes of Pb (dark) and two quarter tubes of Sn (bright)), (**b**) after 1/4 turn, (**c**) after 1/2 turn, and (**d**) after eight turns of t-HPS where a partial enlargement of a section in a red box are shown in (**e**) by SEM. (Reprinted from [47], Copyright 2016, with permission from Elsevier)

t-HPS, where a partial enlargement of a section in the red box is shown by SEM in (e) [47]. It is apparent that there is an excellent bonding between Pb and Sn and the multilayered Pb–Sn laminates were successfully produced by t-HPS processing in a single step process.

4.2 Functionalities of Hybrid Systems

This section focuses on the mechanical properties and additional functionalities of the bulk hybrid nanostructured alloys produced by synthesis methods and techniques which are explained in Sect. 4.1.

4.2.1 Enhanced Properties of Bulk Hybrid Systems Synthesized by Powder Consolidation

Application of SPD for powder consolidation is promising for synthesizing hybrid nanocrystalline systems, but the conditions of initial powders before consolidation by HPT processing can vary the defect structure, leading to different values of hardness and strength in the synthesized bulk metals and alloys. For instance, an example was shown in a study using gas-atomized Co powders having <44 μm particle sizes (based on 325 mesh) without ball-milling and with ball-milling for up to 45 h prior to powder consolidation by HPT for 5 turns under 6 GPa [9]. The initial gas-atomized Co powders having a single *f.c.c.* structure showed allotrope transformation almost completely to an *h.c.p.* structure after HPT, while additional ball-milling for 45 h prior to HPT created a dual structure of *f.c.c.* + *h.c.p.* and the following HPT

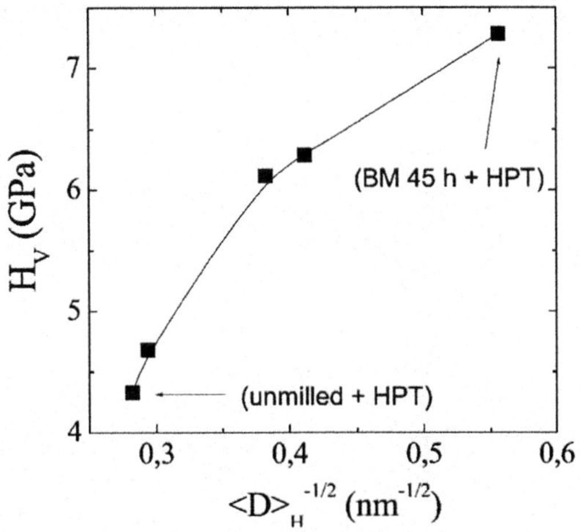

Fig. 4.16 The hardness values of the bulk Co with the inverse square root of the crystallite size of *h.c.p.*-Co. (Reprinted from [9], Copyright 2003, with permission from Elsevier)

processing further increased the phase fraction of *h.c.p.*-Co. Both ball-milling and HPT processing reduce the crystallite size of the *h.c.p.*-Co to ~10 nm, while the different faults of twin faults and stacking faults, especially deformation faults, tend to increase after ball-milling of Co powders and after ball-milling followed by HPT, respectively. Figure 4.16 shows the hardness values of the bulk Co with the inverse square root of the crystallite size of the *h.c.p.*-Co phase [9]. Although these structural changes lead to different types and populations of defects, significantly high hardness of >7.3 GPa was observed for the bulk Co after ball-milling, followed by HPT. This is by contrast to the hardness of <4.5 GPa for the unmilled bulk Co and the difference is attributed to the smaller crystallite size achieved by ball-milling of the Co powders before HPT consolidation.

The consistent trend of improved hardness after ball-milling prior to HPT consolidation is shown by comparing the bulk samples from Ti powders without ball-milling, while both with and without ball-milled powders after HPT show higher yield strength than an HPT-processed bulk Ti [66]. Thus, ball-milling creating pre-straining and/or contaminations may lead to an enhancement in strength of the HPT-consolidated powder materials.

Besides ball-milling, oxidation of pure metal powders can also promote the strength of the HPT-consolidated bulk nanocrystalline materials. Figure 4.17 shows the hardness variation after HPT processing with increasing equivalent strain for (1) as-received pure Ni powders; (2) pure Ni powders with annealing at 400 °C for 10 min in air to create a thin oxide layer on the particle surface, and bulk Ni samples processed at (3) room temperature and (4) −196 °C [35]. A comparison of the saturated high hardness values shows much higher hardness in the bulk nanocrystalline Ni made of powders, and the powders with oxidation demonstrate extremely high saturation hardness of 6.8 GPa. This study also showed an excellent thermal stability of the HPT-consolidated oxidized Ni powders by maintaining the high hardness even after annealing at 418 °C for 1 h in comparison with the HPT-consolidated Ni powders without oxidation that lost the hardness after annealing at >250 °C for 1 h [35].

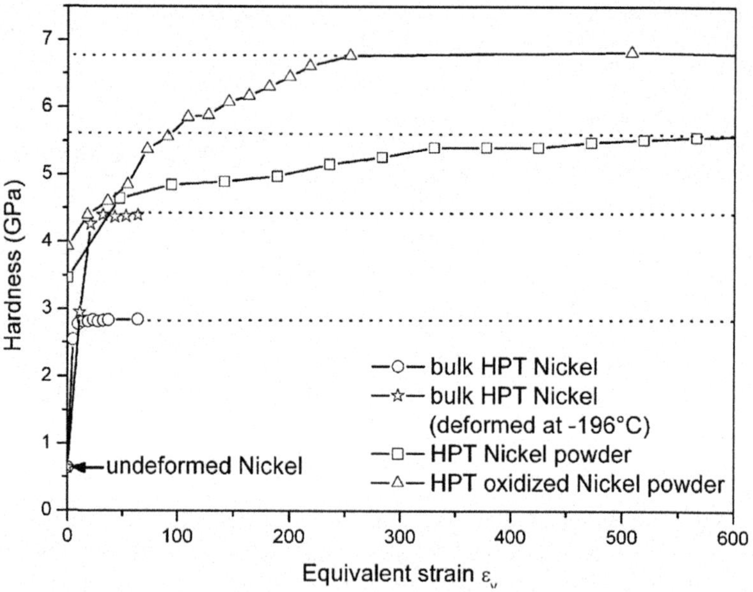

Fig. 4.17 Variation of microhardness of an HPT bulk nickel deformed at room temperature and at ~196 °C and of an HPT-consolidated samples of nickel powders and oxidized nickel powders as a function of the applied equivalent strain by HPT. The hardness of bulk nickel without any deformation is 0.64 GPa (indicated by an arrow). (Reprinted from [35], Copyright 2009, with permission from Elsevier)

Considering the principles of HPT processing, it is reasonable that the high compressive pressure of HPT provides better consolidation of powders with grain refinement, thereby introducing an enhancement in hardness for the resultant bulk sample [4, 66]. In contrast, for understanding the significance of the numbers of HPT turns on the powder consolidation, a mixture of pure Al and Fe powders was consolidated by HPT under 6 GPa for up to 75 turns at room temperature, where the powder mixture forms a nominal composition of Al–10 wt.%Fe with a particle size of ~53 μm [7]. A relative density of >99% was achieved after HPT for 1 turn in the consolidated Al–Fe alloy and an average grain size of ~145 nm was achieved after 75 HPT turns. Figure 4.18 shows stress–strain curves taken in tension at a strain rate of 3.0×10^{-3} s^{-1} at room temperature for the Al–Fe alloy after HPT for different numbers of HPT turns [7]. There is a large improvement in ductility to the maximum elongation to failure of 55% after 5 turns. The high ductility in the early stage of HPT up to 5 turns was due to the significant grain refinement, but with many grains in a dislocation-free state enabling the capability of strain hardening in the bulk alloy. However, further work hardening by dislocation pile-ups at supersaturated Fe solutes in the matrix phase and Fe-rich dissolute particles led to a reduction in ductility with increasing HPT turns. It is important to note that two sets of scattered stress–strain curves are shown for the alloy after 50 and 70 HPT turns and these scatters are due to the inevitable heterogeneity in the composition of the mechanically-mixed Al–Fe material by HPT.

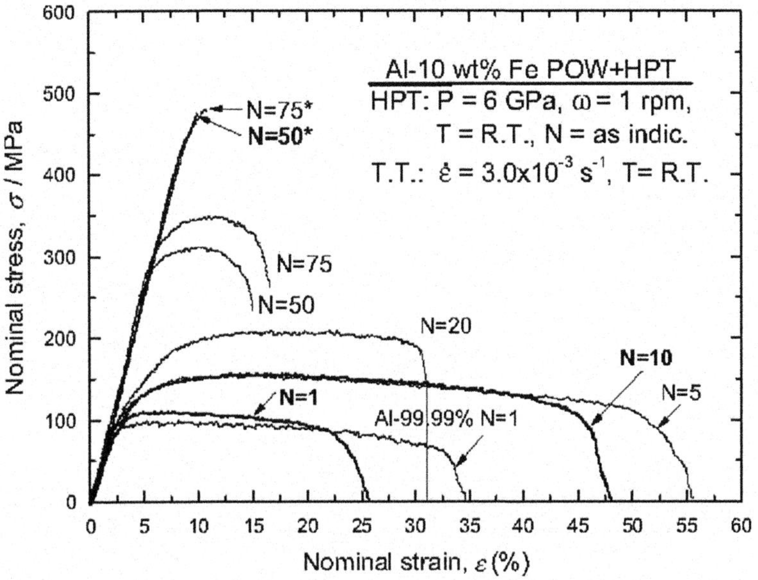

Fig. 4.18 Stress–strain curves of the Al-Fe alloy after HPT processing for different numbers of turns. (Reprinted from [7], Copyright 2012, with permission from Elsevier)

As discussed in Sect. 4.2.1, powder consolidation by HPT enables the synthesis of bulk nanocomposites by the nucleation of metastable phases, including intermetallic compounds and supersaturated solid solutions. It significantly improves the mechanical properties as well as thermal stability. Powders of pure Al and Ni as 50 mol.% of the whole, having particle sizes of <75 μm and <150 μm, respectively, were mixed and consolidated by HPT at 573 K for 3, 10, 50, and 120 turns (equivalent to shear strains of 80, 270, 1700, and 3300, respectively) [67]. Mechanical mixing and grain refinement by HPT introduced first an Al_3Ni intermetallic phase in an early stage of straining to a shear strain of 80 and it transformed to an Al_3Ni_2 intermetallic phase with further straining toward a shear strain of ~1400. The bulk intermetallic-based nanocomposite disk of Ni–Al_3Ni_2 after 50 HPT turns was annealed at 673 K for 24 h and further transformed to a bulk AlNi intermetallic compound. Figure 4.19 shows the variation of microhardness versus shear strain by HPT for the powder-consolidated bulk Al–Ni system, where the plot includes the reference hardness values of as-cast AlNi, and as-HPT Al and Ni processed at 298 K [67]. The dramatic increase in hardness after shear strains of >100 is attributed to the formation of an Al_3Ni_2 phase and the high hardness is much higher than these reference materials. Moreover, excellent thermal stability was observed in the bulk nanostructured AlNi after annealing by reasonably maintaining the high hardness. The result is supported by the maintenance of the measured grain size of ~28 nm after HPT to ~50 nm after annealing. Such high hardness and yield strength due to the formation of TiAl intermetallic compounds are also observed in a study

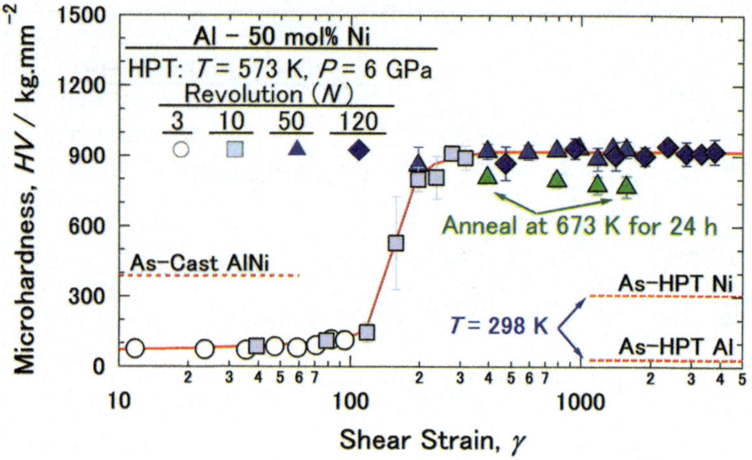

Fig. 4.19 Variation of microhardness versus shear strain by HPT for the powder-consolidated bulk Al-Ni system and after annealing at 673 K for 24 hours. The plot includes some reference hardness values of as-cast AlNi and as-HPT Al and Ni processed at 298 K. (Reprinted from [67], Copyright 2012, with permission from Elsevier)

consolidating an Al–50 mol.% Ti micro-powder by HPT [68]. Annealing of the HPT-consolidated nanostructured TiAl intermetallics demonstrated an improved ductility in compression and it was due to the formation of nanotwins during annealing.

HPT processing for the consolidation of oxide powders was studied recently for the synthesis of bulk functional nanostructured materials. The attractiveness of the HPT process on oxide powders is the formation of high oxygen vacancies influencing the electronic structure [69] of the consolidated bulk nanocrystalline oxides. Moreover, the increase of oxygen vacancies leads to color changes of the HPT-consolidated oxides to dark. Together with phase transformation, these changes in electronic structures and color imply a significant potential in tailoring the electrical and optical properties of the bulk nanocrystalline oxides from powders by HPT consolidation. An example of a color change is shown in Fig. 4.20 for a titanium oxide powder having a color of white consolidated by HPT under 6 GPa at room temperature for up to 4 turns [70]. There is a phase transformation from Anatase TiO_2 to TiO_2-II and gradual changes in color from white observed immediately after HPT compression toward darker green with increasing shear strain by HPT. The color change was due to the formation of oxygen vacancies in the consolidated bulk sample.

HPT processing for 3 turns under 3 and 6 GPa at 25 °C was applied for consolidation of ZnO powders (Wurtzite having an hexagonal structure) to examine the variation in a visible-light photocatalytic activity [71]. A structural change was found after HPT to Rocksalt-ZnO having a cubic structure. The band gaps of the initial ZnO powder and after consolidation at different HPT pressures [71] were

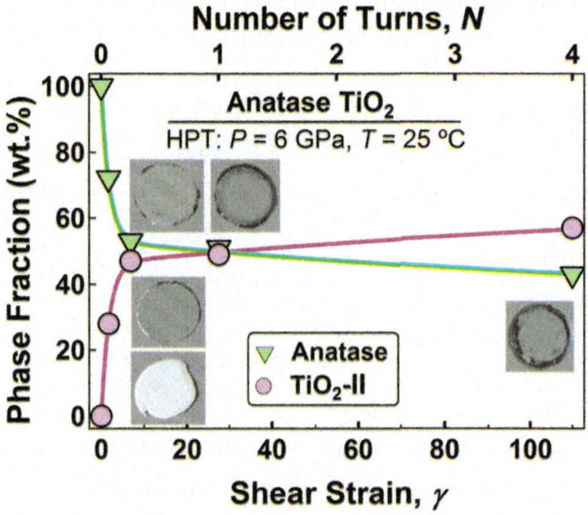

Fig. 4.20 Variations of fractions of Anatase and TiO_2-II as a function of shear strain. (Reprinted from [70], Copyright 2016, with permission from Elsevier)

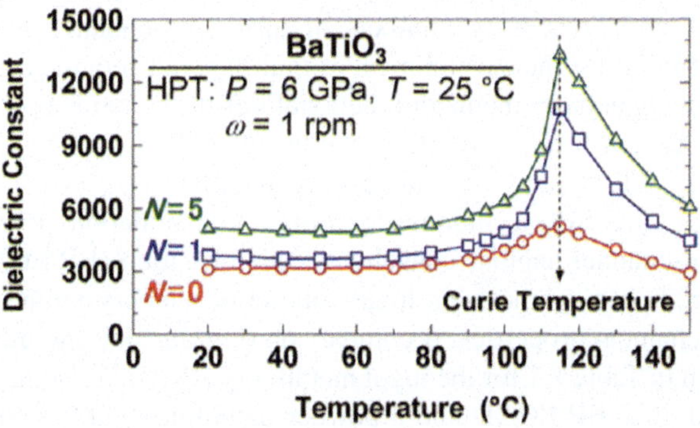

Fig. 4.21 A variation of dielectric constant with temperature for the $BaTiO_3$ powder after HPT for 0 (compression only without torsion), 1 and 5 turns [18]

examined based on the Kubelka-Munk theory [72]. The decrease in the estimated band gap by HPT was observed from 3.1 eV for the initial powder to 2.8 eV and 1.8 eV for the consolidated ZnO under 3 and 6 GPa, respectively. The decrease in band gap is due to an increase of oxygen vacancies in the consolidated ZnO. The degradation rate of Rhodamine B (RhB) under visible light was examined for these three samples [71]. A successful photocatalyst should have a band gap to have a visible-light photocatalytic activity, and thus an excellent photocatalytic activity under sunlight was achieved in the Rocksalt-ZnO having the largest oxygen vacancies after HPT consolidation at 6 GPa, while the initial ZnO powder failed to show any photocatalytic activity under visible light due to its large band gap.

An increasing dielectric constant with increasing numbers of HPT turns from 1 to 5 turns for powder consolidation was observed in a $BaTiO_3$ powder processed under 6 GPa at room temperature. Figure 4.21 shows the variation of dielectric

Table 4.1 Characteristics of the IM material before and after BP-ECAD and PM material after BP-ECAC. (Reprinted from [30], Copyright 2005, with permission from Elsevier)

	IM Al		PM Al
Materials	As-cast	After BP-ECAD	After BP-ECAC
Average grain size (μm)	1600	535	7.2
Density (g/cm^3)	2.70	2.70	2.70
HV (kg/mm^2)	23.5	32.3	52.7
0.2% Proof stress (MPa)	27	60	135
UTS (MPa)	41	64	160
Strain to fracture (%)	58	23	11

constant with temperature for the $BaTiO_3$ powder after HPT for 0 (compression only without torsion), 1, and 5 turns [18]. The increase in the dielectric constant is attributed mainly to a decrease in tetragonality by demonstrating tetragonal-to-cubic structural transformation in the $BaTiO_3$ during consolidation and grain refinement by HPT with increasing numbers of turns. An enhancement in dielectric constant is essential for the development of multilayered ceramic capacitors. These examples showing enhancements in functionalities of the oxide powders after HPT consolidation demonstrate the feasibility of using HPT processing for the synthesis of bulk nanocrystalline functional and energy materials.

Considering other SPD techniques for powder consolidation, processing by back pressure equal-channel angular consolidation/deformation (BP-ECAC/ECAD) for powder materials can synthesize a large volume of nanocrystalline materials with enhanced mechanical properties. A comparison of grain size and mechanical properties is shown in Table 4.1 for the ingot metallurgy (IM) material of an as-cast pure Al before and after BP-ECAD and a powder metallurgy (PM) sample of pure Al powders processed by BP-ECAC for 1 pass at 100 °C with a back pressure of 50 MPa [30]. The initial powders had 90% of particles having a particle size of <60 μm and included 0.38 wt.% of oxygen forming an Al_2O_3 oxide. Because of the fine structure of the initial powders as well as the aid of the oxide, the PM Al after BP-ECAC showed a much smaller grain size than the IM Al after BP-ECAD. The fine microstructure of the PM Al enabled twice or more enhanced 0.2% yield strength and UTS, but less than half of the strain to fracture than the IM Al after BP-ECAD.

Improvement in strength of bulk hybrid materials made of Cu powders with 1 vol.% of carbon nanotube (CNT) was demonstrated after ECAP consolidation by examining a change in the required loads for a fixed spherical indentation displacement of 300 μm. Figure 4.22 shows the load-displacement curves for the Cu powders without CNT and with 1 vol.% CNT after ECAP for 4 and 8 passes [27]. The curves showing higher peak loads imply the high yield strength of materials. Thus, the bulk hybrid materials of Cu-1 vol.% CNT exhibited higher strength than the bulk Cu samples without CNT after ECAP consolidation for both 4 and 8 passes. The high strength is due to CNT reinforcement in the Cu matrix and the resultant material is thus high strength CNT-metal matrix composites.

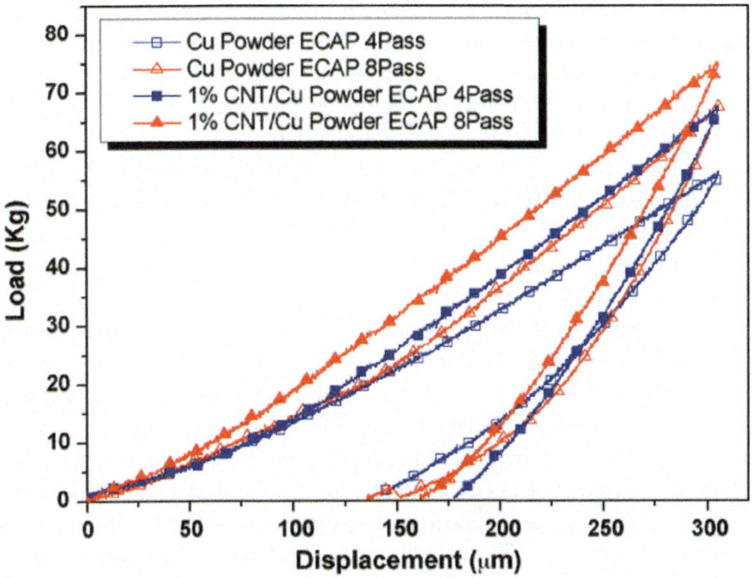

Fig. 4.22 Load-displacement curves for the Cu powders without CNT and with 1 vol.% CNT after ECAP for four and eight passes. (Reprinted from [27], Copyright 2007, with permission from Elsevier)

4.2.2 Enhanced Properties of Bulk Multilayered Hybrid Systems

The multilayered hybrid systems after SPD processing, such as ARB and ECAP, generally have a large sample volume which permits the testing of their mechanical properties in multilength scales. The macro-scale tensile testing and even nano-scale nanoindentation can quantify the mechanical behavior in stress–strain curves providing Young's modulus, strength, and the ductility of materials. In addition to these general testing approaches, testing by micropillar compression provides the plastic behavior from an elastic deformation region to failure involving much smaller sample volumes. This testing technique was applied to examine a limited volume of a nanolayered Cu/Nb hybrid system processed by ARB [53, 73]. The ARB-processed Cu/Nb showed a UTS of 1.7 GPa and a strain to failure in compression of 0.1 with a limited strain softening. The study analyzed the mechanism of deformation to failure by diffuse shear banding which involves the homogeneous distribution of many fine shear bands localized within Cu grains forming micro-bands along the length of the micropillar. It should be noted that the failure mode creates visible slip-bands on the micropillar surfaces during slip, but it does not involve sliding at the Cu/Nb interfaces. A series of bulk-scale tensile tests on the nanolayered Cu/Nb hybrid material revealed that the thinner layer thicknesses within the bulk system increase its UTS during deformation when compared with the alloy system having different layer thicknesses from 58 nm to 2.4 μm.

By contrast, when conventional rolling is conducted for introducing a multilayered laminate system with inconsistent thicknesses of ~200 μm–1.0 mm of each layer, the multilayered alloy after rolling shows anisotropy in the plastic deformation behavior. The Ti/Al/Mg/Al/Ti laminates processed by ARB having gauge lengths oriented along the rolling direction (RD), the transverse direction (TD), and a 45° angle toward RD were tested in tension at room temperature [51]. The tensile gauge length parallel to RD showed the largest UTS of 434 GPa and the gauge with a 45° angle shows a similar value, while the gauge along TD showed the lowest UTS as well as less elongation to failure of the laminates. A main factor contributing to the anisotropy in the deformation behavior of the laminates is the Mg layer, which shows complex microstructures involving coarse grains, twins, deformation bands, a shear zone, and dynamic-recrystallized grains when the laminates are deformed along RD, while no twins appear and other microstructural features are less in the Mg layer when deformed along TD. Moreover, there are four stages of plastic deformation. First, deformation occurs in the elastic and plastic fields toward obtaining the maximum stress. Second, the flow stress drops due to the segregation of Mg/Al interface and this is followed by a third stage where the fracture of the Mg layer occurs. Fourth, the Ti layer fractures and the hybrid laminates fracture ultimately.

The geometry of the multilayers, including the thickness of the layers, influences the properties of the bulk hybrid materials. A bulk nanocrystalline bimetallic rod of AA6201 Al alloy/316 L Steel processed by ECAP was tested for examining the influence of the processing history and the bimetallic layer thickness on the overall strength and electrical conductivity [57]. The variations of yield strength and electrical conductivity were described against the volume fraction of steel in the as-received AA6201/steel rods having 10 mm diameter. Specifically, three separate AA6201/Steel bimetallic rods were prepared with different steel diameters of 3, 4, and 5 mm at the cores (Ø3, Ø4, and Ø5 steel core) surrounded by AA6201 tubes forming 10 mm diameter rods and processed for 0, 1, and 4 ECAP passes at 175 °C. The yield strength increases with increasing numbers of ECAP passes and the tendency is highest with the thicker Steel core, while the electrical conductivity goes down with increasing amounts of steel within the sample regardless of the ECAP processing. Since the density increases significantly with the increasing volume of steel in the bimetallic Al/Steel rods, it is concluded that the thinnest steel in the bimetallic rod demonstrates reasonably high yield strength and thereby attains the highest specific strength while maintaining the highest electrical conductivity.

4.2.3 Enhanced Properties of the HPT-Processed Hybrid Metals and Composites

Bulk-state reactions of dissimilar metals by HPT were demonstrated in a variety of combinations of metals and alloys (see Sect. 2.2.2). Some combinations of dissimilar metal sets, such as Al/Ti/Al and Al/Fe/Al systems, demonstrated successful bonding without the notable dissolution of elements and the formation of intermetallic phases even after HPT for 50 turns under 6 GPa at room temperature [74]. This is in contrast with the powder consolidation of the same element combinations by HPT under consistent processing conditions, where the processed hybrid nanomaterials involve intermetallic phases. Thus, processing by HPT synthesizes very different final hybrid nanostructured alloys through the bulk-state reactions and the powder metallurgy approach. After high numbers of HPT turns, often more than 80 or higher, the bulk-state reactions of dissimilar metals by HPT involve the formation of metastable phases by accelerated diffusion of atoms under the deformation mechanisms explained earlier (Sect. 4.2.2).

The bulk-state reactions of dissimilar metals and alloys by HPT for 10–20 turns demonstrate the bonding of the base metals with a typical hardness development within the base metals, so that the hardness is low at and around the disk center and becomes higher toward the disk edge. By the accelerated diffusion of the atoms by continuous severe plastic deformation to higher numbers of turns, the formation of metastable phases develops at the disk edge and it often provides extreme hardness values which are higher than the values achieved by the separate base materials after grain refinement. Further additional turns extend the metastable phases toward the disk center and ultimately the extreme hardness becomes distributed homogeneously, while any further phase transformation may cause additional hardness change with increasing numbers of HPT turns.

An example of a representative hardness development during a bulk-state reaction by HPT is shown in Fig. 4.23 displaying the color-coded contour hardness maps for the disk cross-sections of the Nb–Zr system processed with the metal stacking of Nb/Zr/Nb under 6 GPa for up to 100 turns at room temperature [75]. For the Nb–Zr system, 100 turns are not sufficient to introduce a homogeneously distributed high hardness across the disk diameter. Nevertheless, the results show the development of extreme hardness at the disk edge with improved hardness at the disk center after 50–100 turns by the introduction of a true nano-scale microstructure with grain sizes of 20–40 nm as well as the formation of twinning in the complete mixture of Nb and Zr.

One of the important properties that the bulk-state reactions by HPT have improved significantly is the strength-to-weight ratio by strengthening the materials while simultaneously reducing the density through the introduction of HPT-synthesized metal matrix nanocomposites by the bonding of dissimilar metals that include one or more lightweight metals and alloys. Some examples are shown in Table 4.2, where the densities and microhardness values of the Al–Mg, Al–Cu, and Al–Fe systems synthesized by the solid-state reactions are measured at the disk edges after different numbers of HPT turns under 6 GPa and these values are used for estimating their strength-to-weight ratios. For comparison purposes, the density, hardness, and strength-to-weight ratios are calculated for the HPT-deformed base metals and alloys without bulk-state reaction. There is a gradual decrease in density by the accelerated diffusion of these elements leading to the formation of intermetallic compounds which introduce a significant increase in hardness besides grain refinement with increasing numbers of HPT turns. Subsequently, the

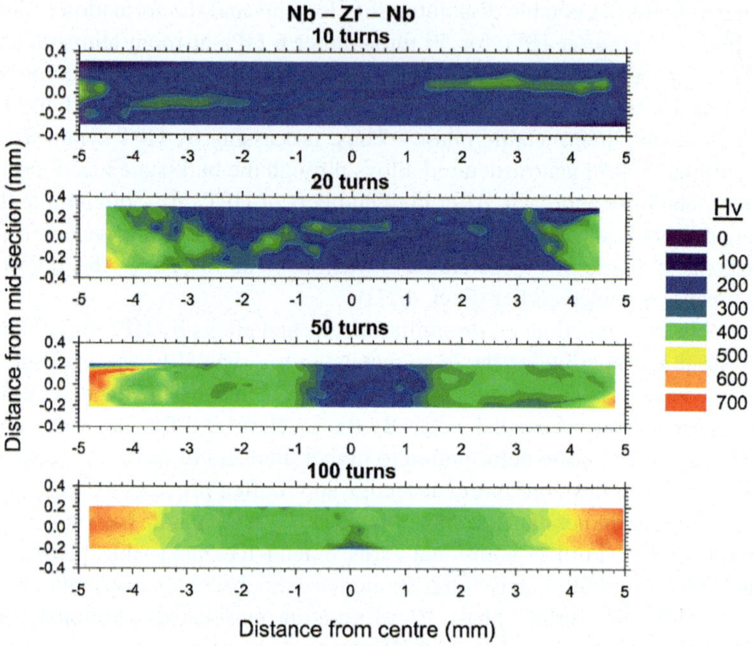

Fig. 4.23 The color-coded contour hardness maps for the disk cross-sections of the Nb-Zr system processed on the metal stacking of Nb/Zr/Nb under 6 GPa for up to 100 turns at room temperature. (Reprinted from [75], Copyright 2020, with permission from Elsevier)

strength-to-weight ratios are enhanced dramatically in these HPT-synthesized alloy systems and similar enhancements are not achievable for the base metals when they are processed by HPT without the bulk-state reactions. The results imply the significance of the processing approach for the development of lightweight hybrid nanostructured alloys from conventional simple engineering metals and alloys.

In an early stage of the bulk-state reactions of dissimilar metals by HPT, there is a mixture of refined phases of base metals and a small content of nucleated intermetallic compounds. These different phases show very different mechanical properties in a local region of the alloy system, thereby demonstrating plastic instability. For example, the Al–Mg system shown in Table 4.2 which was synthesized by the bonding of separate Al and Mg disks by HPT for 5 turns demonstrated very scattered load-displacement curves for 15 separate measurements by nanoindentation taken at the disk edge [76]. Such non-homogeneous behavior of the mechanically bonded materials by HPT disappeared after 10 or higher numbers of turns. A recent report on an Al–Cu system bonded by HPT for up to 200 turns under 6 GPa-conducted tensile testing and Fig. 4.24 show the stress–strain curves for the Al–Cu system with the base materials [77]. The mechanical behavior of the Al–Cu system shows an expected high yield strength with increasing HPT turns, while the ductility of the Al–Cu system even after 150 and 200 turns was similar to a base metal of the Al 1050 after HPT for 10 turns. Such reasonable ductility was expected by the improved

Table 4.2 The measured density, the maximum Vickers hardness value, and the estimated strength-to-weight ratio for the solid-state bonded Al–Mg system after HPT for 5–20 turns, Al–Cu system after 20 and 60 HPT turns and Al–Fe system after HPT for 20 turns with the reference separate materials of Al-1050 and ZK60 alloys after HPT for 5 turns and commercial purity (CP) Cu and CP Fe after HPT for 10 turns [61]

Solid-state bonded alloy system	Density ($g\ cm^{-3}$)	Maximum hardness (Hv)	Strength-to-weight ratio ($MPa\ cm^3\ g^{-1}$)
Al/Mg/Al (HPT: 5 turns)	2.50	130	170
Al/Mg/Al (HPT: 10 turns)	2.48	270	350
Al/Mg/Al (HPT: 20 turns)	2.34	330	455
Al/Cu/Al (HPT: 20 turns)	4.47	400	290
Al/Cu/Al (HPT: 60 turns)	4.32	500	390
Al/Fe/Al (HPT: 20 turns)	3.81	320	296
Commercial alloy			
Al-1050 (HPT: 5 turns)	2.73	65	80
ZK60 (HPT: 5 turns)	1.84	110	190
CP Cu (HPT: 10 turns)	8.96	150	55
CP Fe (HPT: 10 turns)	7.87	310	128

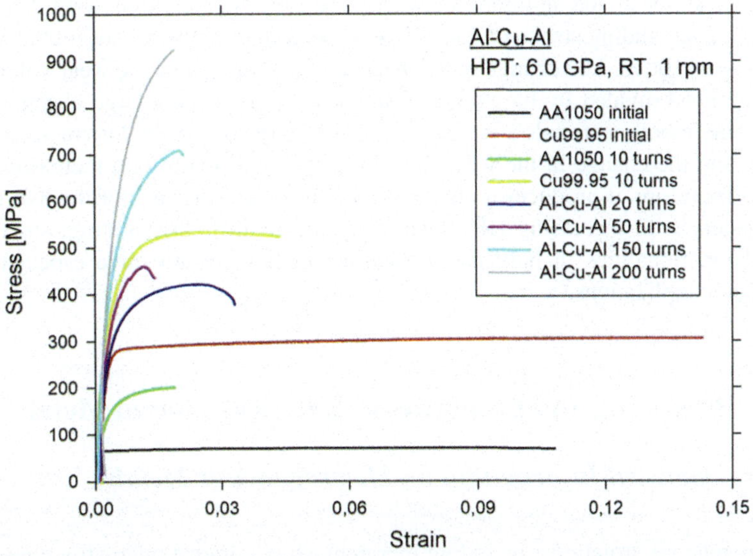

Fig. 4.24 Typical tensile engineering stress–strain curves for the mechanically bonded Al-Cu system after HPT for 20, 50, 150, and 200 turns together with curves for the base metals of Al-1050 and 99.95Cu alloys with and without HPT. (Reprinted from [77], Copyright 2020, with permission from Elsevier)

plasticity estimated by the improved strain rate sensitivity in the mechanically-bonded Al–Cu system after HPT with increasing numbers of turns [61].

The inhomogeneous distributions of the nucleated nano-scale intermetallic compound phases within the refined metal matrix phases influence the corrosion properties of the mechanically-bonded hybrid nanocrystalline alloys processed by HPT. An example was shown in the Zn–Mg system where stacked disks of Zn/Mg/Zn were bonded by HPT for 15 and 30 turns under 6 GPa [78]. In practice, HPT processing through 15 turns for mechanical bonding of Zn and Mg disks showed a tendency of lower corrosion resistance, and therefore a faster corrosion rate and more negative corrosion potential, than the alloy system after 30 turns. The result was attributed to (a) more homogeneously refined and distributed Zn and Mg phases after 30 turns, where, in general, a grain size reduction improves the corrosion resistance in NaCl electrolytes, and (b) the development of a homogeneous multilayered nanostructure after 30 HPT turns delaying the reactions of the synthesized Mg_2Zn_{11} and $MgZn_2$ intermetallics that usually promote the formation of micro-galvanic cells which ultimately trigger a faster dissolution of the passive layer of ZnO.

The strengthening of engineering metals often (but not always) improves the wear properties by showing either slower wear rates under a specific applied load or a higher applied load under a specific wear rate during tribology tests. A hybrid Al–Cu system synthesized by HPT for 20 and 60 turns shows, as is apparent in Table 4.2, an improved hardness and strength-to-weight ratio by the severe mixture of Al and Cu, leading to the formation of hybrid nanocrystalline metal matrix nanocomposites. A series of tribology tests on Al–Cu metal matrix nanocomposites after 60 turns demonstrated an improved wear resistance when the tests were conducted under a fixed sliding stroke of 2.0 mm at a speed of 0.002 m s^{-1} for a total sliding distance of 100 m. The change in the wear rate is computed by the wear volume (or volume loss) divided by the sliding distance [79]. There is a large change in the wear rate in both materials after 20 and 60 HPT turns due to the loss of wear resistance, and the change in the wear rate shifts to higher loads with increasing HPT turns. Therefore, it produces an improvement in wear resistance of the hybrid Al–Cu system with increasing HPT turns. Together with lightweighting, such additional functionalities in bulk nanocrystalline metals are attractive especially for structural applications.

4.3 Processing and Properties of Heterostructured Metals

4.3.1 General Information on Heterogeneous Metals

Heterogeneous structures, including hierarchical and graded microstructures, have evolved in non-metallic materials over millions of years through natural selection and optimization, such as in shells, bones, and plant tissues, where structural and compositional changes occur gradually throughout the material [80, 81]. Introducing gradient structures into metallic materials is also a method for optimizing the physical and mechanical performance while minimizing cost. Specifically, heterostructured metallic materials cover a variety of structural designs of grains and phases, as in bi-modal structured materials, dual-phase materials, gradient structured materials, harmonic structured materials, hetero-lamella structured materials, laminate structured materials, and metal matrix composites. Thus, many hybrid materials synthesized through mechanical bonding of dissimilar metals (Sect. 4.1) often demonstrate a heterogeneous microstructure consisting of different phases.

The finely developed gradient structure, including a hierarchical structure at the atomic scale [82], in engineered metals demonstrates unique extra strain hardening, leading to both high ductility and toughness [83]. In practice, when a material with a graded grain size and intra-phase gradients is under tension, the onset of plastic deformation initially occurs in the coarse grains and gradually propagates into the smaller grains with increasing load. The plastic deformation releases intergranular stress between neighboring grains of different sizes so that strain localization is suppressed [84, 85]. Such extraordinary strain hardening that maintains high ductility, is an inherent feature of gradient structures and does not exist in homogeneous materials. It also provides a new and important strategy for developing materials that are both strong and ductile.

The benefits of graded and hierarchical phase distributions in producing metals with dramatic improvements in strength and ductility were demonstrated in gradient-structure/coarse-grains (GS/CG) steels [85]. The strain delocalization of gradient microstructures enhances fatigue resistance during cyclic loading [86]. A heterogeneous and hierarchical structure also imparts excellent mechanical properties, especially strength and ductility, to various nanocrystalline metal systems. Figure 4.25 shows a plot of normalized yield strength against uniform tensile strain for various

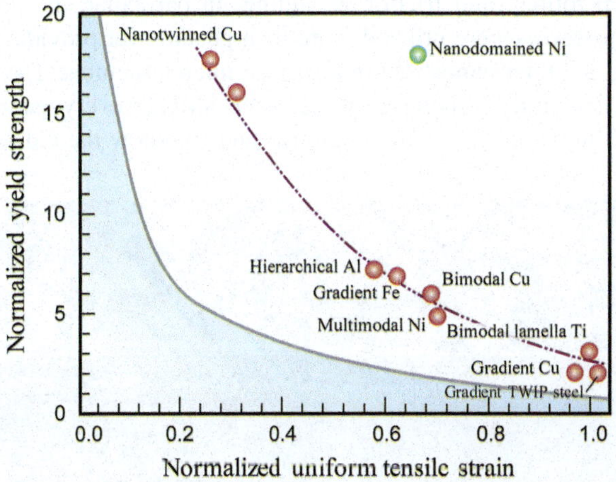

Fig. 4.25 Ductility vs. yield strength of gradient-structure/coarse-grained (CS/GS) steel (open circles) compared to similar interstitial free (IF)-steels with more homogeneous microstructures. (Reprinted from [87], Copyright 2017, with permission from Elsevier)

metals having different types of heterogeneous microstructures [87]. The shaded area under the negative exponential curve delineates the strength–ductility relationship for metals and alloys having a range of homogeneous grain sizes showing the paradox of strength and ductility [88]. It has been one of the major challenges of materials science to overcome this paradox in the processing of nanostructured metals and materials and this plot demonstrates the feasibility of microstructural design through the use of a series of processing procedures toward a strength–ductility synergy.

In this specific topic of heterostructured metals, it is important to understand the relationship between heterogeneous and gradient microstructures and their properties quantitatively for developing the global properties of the overall heterostructured materials. The development of processing techniques for such unique microstructures has been challenging, but several SPD techniques demonstrated excellent contributions to the synthesis of bulk heterostructured metallic materials. The following section provides several heterostructured metals processed by surface mechanical treatments, conventional thermomechanical processing and different SPD techniques except mechanical bonding of dissimilar metals (see Sect. 4.1) and then discusses their improved mechanical properties and functionalities. Moreover, the architecturing of microstructure and a strategy of material design by SPD are also discussed.

4.3.2 *Processing of Heterostructured Metals*

A heterogeneous/gradient structure is generally produced by shot peening, surface mechanical attrition treatment (SMAT), surface mechanical grinding treatment (SMGT), deep rolling, and friction & sliding. In particular, the SMAT [89] and SMGT [90] processes were utilized more than peening for providing smother surface finishing and finer nanostructured surface microstructures. The representative resultant gradient microstructures of Cu after SMGT at cryogenic temperature (~173 K) are shown in Fig. 4.26, where (a) and (b) show the Cu samples having

Fig. 4.26 Representative microstructure of SMGT-treated Cu samples having different volume fractions of gradient nano-grained layers of (**a**) about 4.3% and (**b**) 13.0%. (Reprinted from [91], Copyright 2019, with permission from Elsevier)

different volume fractions of gradient nano-grained (GNG) layers of about 4.3% and 13.0%, respectively [91]. Another study describes the surface mechanical treatment which introduces a GNG structure with grain sizes of 20–100 nm from the gauge surface to 60 mm thick depth toward the core and grain sizes of 100–300 nm at a depth of 60–150 mm [84]. These GNG structures are followed by a deformed coarse-grained (CG) microstructure in a layer at 500–700 mm away from the sample surface. Moreover, the surface nanocrystallization procedure of SMAT was used for a new approach to enhance the global mechanical properties of materials by optimizing the surface structure. A nanocrystalline surface layer with high thermal stability was generated in a pure Al after SMAT and the yield strength of samples was enhanced to about two times higher than their coarse-grained counterpart [92].

Processing of heterostructured metals is available on a large length scale through the application of conventional cold-rolling techniques and several SPD techniques. A heterogeneous lamella structure (HLS) was observed in brass (Cu–30% Zn alloy) after cold-rolling for 80% thickness reduction (R80) and partial recrystallization by applying a post-deformation annealing (PDA) treatment at 260 °C for short-time periods of up to 30 min [93]. The synthesis of HLS materials utilizes a typical heterogeneous microstructure after cold-rolling involving some fields with elongated fine grains and other regions with highly deformed ultrafine grains having a high dislocation density leading to dislocation cells and subgrains including deformation twins. After a short time of PDA, recrystallized grains with high growth rate behave as a soft domain in the final HLS microstructure. By contrast, the remainder is recognized as a hard domain composed of two different types of grains: elongated fine grains after cold-rolling that failed to recrystallize and maintained sharp grain boundaries after PDA, and newly recrystallized UFG grains after PDA. This study demonstrated an optimized combination of excellent strength and ductility in the HLS brass after cold-rolling and ~20 min PDA and it was attributed to hetero-deformation induced (HDI) strengthening with improved work hardening. The excellent combination of strength and ductility and the enhanced strain hardening ratio in the HLS brass after cold-rolling and PDA are shown in Fig. 4.27a, b,

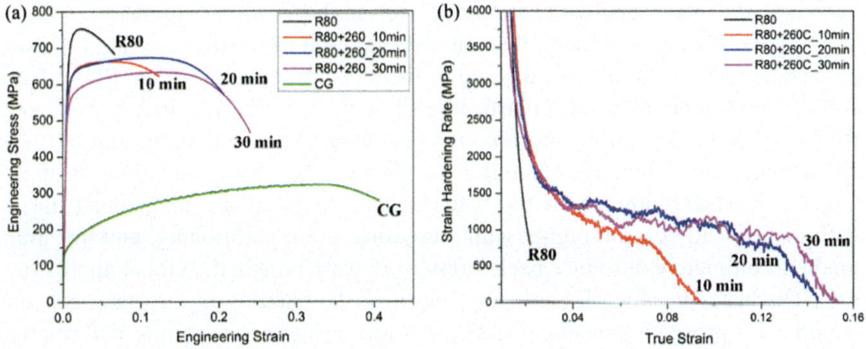

Fig. 4.27 (**a**) Strain-stress curves and (**b**) strain-hardening rate curves for brass after cold-rolling for 80% thickness reduction (R80) and partially recrystallized at 260 °C for different time periods. (Reprinted from [93], Copyright 2020, with permission from Elsevier)

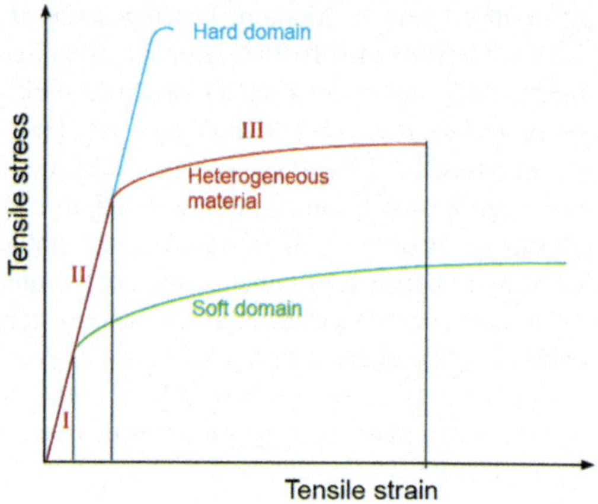

Fig. 4.28 The general stress–strain behavior involving the soft and hard domains of the heterostructured material during tensile deformation, where there are three separate deformation stages in the stress–strain curve of the heterostructured material [94]

respectively [93]. These improvements are influenced by the volume fractions of soft and hard domains, the thickness/spacing of soft domains and the misorientations of domains.

It is important to understand the general deformation mechanism involving the soft and hard domains of the heterostructured materials during plastic deformation. The general mechanical behaviors of these domains and a global heterogeneous material under tension are schematically illustrated in Fig. 4.28, where there are three separate deformation stages in the stress–strain curve of the heterostructured material [94]. In stage I, both soft and hard domains deform elastically within the heterogeneous microstructure. In stage II, the soft domains start to produce plastic strain by dislocation slip while the hard domains remain elastic, thereby creating a mechanical incompatibility. Since the soft domain can accommodate more strains during plastic deformation, a strain gradation accommodated by geometrically necessary dislocations can be formed near the domain interface in order for the domain interface to be continuous under deformation. While the soft domain surrounded by the hard domain matrix cannot change its shape until the hard domain matrix starts to deform plastically, the strain gradient can make the softer phase stronger so that geometrically necessary dislocations pile-up at the domain boundaries without crossing the domain boundary, thus building up high back-stresses within the soft domain and leading to a strengthening of the global material. In stage III, both the soft and hard domains deform plastically, while the soft domains that were already strained in stage II maintain much higher strain than the hard domains during stage III, leading to a strain partitioning. Specifically, the strain gradations near the domain interface within both the soft and hard domains are maintained during deformation and become higher with increasing strain partitioning, and the strain gradients ultimately introduce back-stress work hardening in the global microstructure. The back-stress work hardening improves the ductility in the heterogeneous material by preventing necking during tensile testing. This unique deformation mechanism often provides excellent mechanical properties which are beyond the simple linear summation of the predicted properties by the volume fraction-based simple rule-of-mixtures.

Fig. 4.29 Schematic drawing of a heterogeneous lamella structure with elongated soft coarse-grained domains embedded in an ultrafine-grained matrix [94]

A similar synthesis procedure of cold-rolling to a thickness reduction of 50% and a short PDA treatment at 1223 K for 5 min followed by water quenching was applied to a high-entropy alloy, $Fe_{29}Ni_{29}Co_{28}Cu_7Ti_7$, having an *f.c.c.* single phase [95]. The high-entropy alloy maintained the single-phase structure, but introduced a heterogeneous microstructure with a gradation in grain size after annealing. The heterostructured microstructure produced an extraordinary work-hardening ability, so that the high-entropy alloy demonstrated an excellent strength of 614 MPa and an extreme ultimate tensile strength of ~1308 MPa with a significant ductility of ~23% [95]. Moreover, asymmetrical-rolling with PDA can be another processing procedure for the synthesis of an HLS microstructure, where the asymmetrical rolling used a set of rolling mills with 45 mm diameter rolls and these moved at different velocities of 1 m/s and 1.3 m/s [96].

A strategy of cold-rolling followed by PDA together with a material selection synthesized a heterostructured material with an even further complicated HLS structure. Schematic drawing of a heterogeneous lamella structure with elongated soft coarse-grained domains embedded in an ultrafine-grained matrix is shown in Fig. 4.29 [94]. An example was shown in a 316 L stainless steel processed by a conventional cold-rolling for 85% thickness reduction followed by a PDA treatment at 750 °C for 5–25 min [97]. The as-received stainless steel involved a major phase of austenite and a minority phase of martensite, and the cold-rolling introduced a strain-induced martensite transformation. Moreover, the cold-rolling condition produced different phase sizes and volume fractions of nano-grains (NG), nano-twins (NT), and lamellar coarse grains (LCG), and the PDA treatment introduced recrystallized grains (RG). The formation of heterogeneous lamella structures involving these different microstructural components are developed at different microstructural stages of (1) the as-received stainless steel having coarse-grains (CG) with twins, (2) after cold-rolling involving NT bundle surrounded by NGs between

Fig. 4.30 EBSD images of two different HLS of Ni after (**a**) liquid nitrogen rolling for a total thickness reduction of 84% followed by PDA at 500 °C for 13 min, and (**b**) ECAP for a total strain of 8 and liquid nitrogen rolling followed by PDA at 475 °C for 5 min. (Reprinted from [99], Copyright 2020, with permission from Elsevier)

LCGs, (3) after PDA for 10–15 min introducing some RGs, and (4) after PDA for 25 min introducing widely developed RGs with some remaining LCGs. With these structural changes at the different processing stages, the highest strength was achieved after cold-rolling with a limited ductility, but a reasonably high hardness with an improved ductility was observed after PDA due to the formation of HLS involving soft recrystallized domains.

Further developments in processing of HLS by cold-rolling and a short PDA treatment were demonstrated in Cu [98] and Ni [99] by the application of pre-straining the materials by ECAP. The HLS metals synthesized by cold-rolling followed by PDA generally introduce coarse-grained lamellae embedded in a coarse-grained matrix containing a very high density of dislocations, while UFG grains introduced by ECAP before cold-rolling provide the final HLS having coarse-grains in the UFG/nano-grained matrix. Such a difference in the HLS formation is apparent in Fig. 4.30 showing the EBSD images of two different HLS of Ni after (a) liquid nitrogen rolling for a total thickness reduction of 84% followed by PDA at

500 °C for 13 min, and (b) ECAP for a total strain of 8 and liquid nitrogen rolling followed by PDA at 475 °C for 5 min [99]. In the coarse-grained HLS, the matrix having a high density of dislocations strengthens the whole material by behaving as a hard domain, and by contrast, the coarse lamellae are recognized as the soft domain. The HLS having the UFG matrix as shown in Fig. 4.30b demonstrated significant recovery and recrystallization during PDA. The material shows the UFG matrix as a hard domain and the coarse-grains behave as a soft domain. In tensile testing, both strength and elongation to failure were higher in the HLS with coarse grains in the UFG matrix. Specifically, the soft domain demonstrated shear band delocalization and dispersion and the well-developed ultrafine/nano-grained matrix prevented crack formation and shear band localization. The optimized microstructural formation of soft and hard domains was achievable by ECAP and cold-rolling followed by PDA so that both strain hardening capacity and back-stress strengthening were enhanced in the global heterogeneous Ni.

A similar strategy of applying two different microstructural refinement processes for further enhancing the general synergetic strengthening of heterostructured metals was shown in a gradient-structured brass (Cu–30 wt.% Zn) processed by multiple-pass friction stir processing (FSP) and rotationally accelerated shot peening (RASP) of a surface nanocrystalline treatment followed by annealing at 600 °C for 2 h [100]. The processed material demonstrated a higher yield and ultimate strength than only RASP followed by annealing and the initial coarse-grained brass. The main contribution to the improved yield strength was due to the presence of UFG/nano-scale grains and the enhanced dislocation density within the gradient-structured brass. In practice, the enhanced synergetic strengthening of the heterogeneous material contributed to >33% of the total strength of the gradient-structured brass, where the yield strength without the synergetic strengthening enhancing the heterogeneous structures is assumed for the homogeneous materials.

Processing of HLS materials has been further studied to improve the yield and ultimate tensile strengths in engineering metallic materials as well as exploring the HLS manufacturing procedures which must be simple and readily adaptive to future large-scale manufacturing. In this context, a low-carbon steel was applied to a two-step heat treatment followed by warm-rolling for a 90% sample thickness reduction by repetitive passes at 350 °C [101]. Specifically, the two-step annealing was used to prepare a special microstructure before warm-rolling. The first step of heat treatment was conducted at 950 °C for 60 min followed by quenching to form a high density of lath martensite, and the second step used 820 °C for 10 min followed by quenching to obtain an ultrafine fibrous ferrite-martensite dual-phase microstructure. During the warm-rolling at the optimized temperature of 350 °C, deformation compatibility was improved in the ferrite-martensite dual-phase heterostructured alloy by reducing mechanical incompatibility through solid solution strengthening of the ferrite phase by the oversaturated carbon atoms, so that a co-deformation of the two different phases was achieved for a high degree of heterostructure refinement. Moreover, the warm rolling enhanced the stable nano-lamellae by increasing carbon segregation at the lamellar boundaries, where the nano-lamellae are finer than after cold-rolling. The warm-rolled heterostructured steel demonstrated

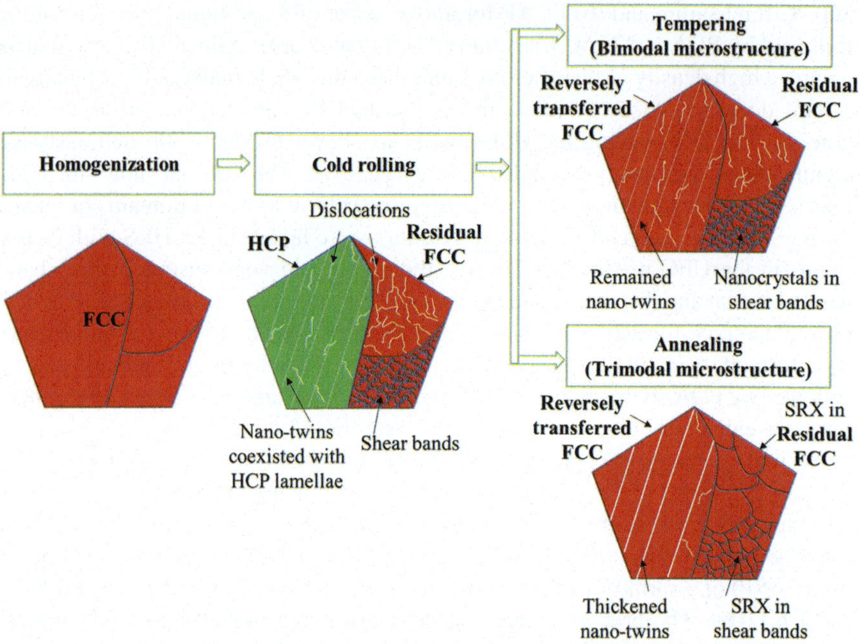

Fig. 4.31 Schematic drawing of multi-phase heterogeneous microstructures produced by thermomechanical processing. (Reprinted from [102], Copyright 2018, with permission from Elsevier)

record-high yield and ultimate tensile strengths of 2.05 GPa and 2.15 GPa, respectively, in the category of low-carbon steels. These high strengths are due to the concurrent contributions of Hall–Petch strengthening from the ultrafine ferrite lamellae, dislocation strengthening from the high intragranular dislocation density, and solute strengthening from the extra carbon solutes diffused from the martensite phase.

It is important to develop economical and sustainable processing technologies that can precisely control and produce the heterostructures in bulk metals and materials [102]. Nevertheless, systematic thermomechanical processing involving cold/warm rolling followed by subsequent annealing provides a positive perspective for developments in the feasible and cost-effective manufacturing procedures for producing heterostructured materials having extreme mechanical properties [101, 102]. Schematic drawing of multiphase heterogeneous microstructures produced by thermomechanical processing is shown in Fig. 4.31 [102].

Finally, there is currently a limited study of heterostructured functional materials despite their excellent potential for various extraordinary functionalities [100]. This is due to the current developmental stage of the processing techniques and procedures for the heterostructured metals, since this leads to an overall lack of understanding of the structure–property relationships and microstructural control in the heterostructured functional materials. As shown in Table 4.3 [103], there are various conflicting properties in engineering materials. Thus, developments in processing of

Table 4.3 Characteristics of the IM material before and after BP-ECAD and PM material after BP-ECAC. (Reprinted from [30], Copyright 2005, with permission from Elsevier)

	Property X	Property X
1	Strength	Ductility
2	Coercivity	Magnetization
3	Electrical conductivity	Seebeck coefficient
4	High electrical conductivity	Low thermal conductivity
5	Reactant mobility	Catalytic active site
6	Polarization	Breakdown strength
7	Ferromagnetism	Ferroelectricity
8	Conductivity	Transparency
9	Conductivity	Strength

heterostructured metals and materials and understanding their structures can overcome the trade-off of the conflicting properties and functionalities, such as the demonstrations of both extended strength and ductility.

4.3.3 Architecturing of Microstructure by Severe Plastic Deformation

Including a development of heterostructured metals, significant recent challenges in Materials Science include the tuning and architecturing of engineering metals which can exhibit extraordinary properties and functionalities. As was shown in Table 4.3, there are numbers of conflicting properties in engineering materials. Conventional alloying of metal elements both by conventional casting and a powder metallurgy approach produces general alloys and composite materials, while the properties of these materials generally follow the rule-of-mixtures. This means the condition is expressed in Fig. 4.32a where the properties of Materials 1 and 2, M_1 and M_2, respectively, reach to either B, C, or D by alloying [104].

Architecturing of unique microstructures in engineering materials, as in the formation of heterogeneous and gradient microstructures, is anticipated to have an excellent potential for exceeding the rule-of-mixtures in the achievable sets of properties and functionalities, as shown in the earlier Sects. 4.3.1 and 4.3.2, and it is demonstrated in Fig. 4.32b, where Materials 1 and 2 can achieve a set of properties reaching to A that is exceptionally higher than M_1 and M_2 as well as B, C, and D in Fig. 4.32a. Figure 4.32c demonstrates an extended condition of (b), where the architectured materials and alloys involve nanostructures so that at least the property M_1 can be extended to N_2 and the ultimate properties of the architectured alloys and composites will reach MS-A which is higher than N_1 and A in (b) [104]. Thus, processing of heterostructured nanocrystalline metals can be suggested to utilize the SPD techniques with/without complementary heat treatments for further development in architecturing microstructures in bulk heterostructured nanocrystals having extraordinary properties and functionalities [105].

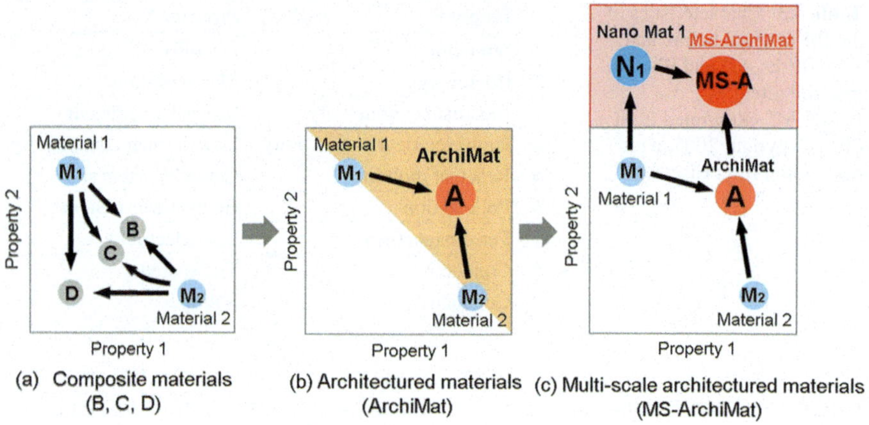

Fig. 4.32 A conceptual framework of multi-scale architecturing: (**a**) composite materials (B–D) made of two materials (M1, M2), (**b**) architectured materials, and (**c**) multi-scale architectured materials (MS-A) made of a nanostructured M1 having a developed property of N1 with M2. The colored regions in (**b**, **c**) represent expanded material spaces by architecturing and nanostructuring. (Reprinted from [104], Copyright 2016, with permission from Elsevier)

It is worth considering a long-term research direction in the fields of SPD processing of bulk hybrid nanostructured materials and architecturing of heterostructures in nanocrystalline metals and alloys. As was shown in the earlier Sects. 4.1 and 4.2, various SPD processing techniques enable the creation of hybrid structures and heterostructures in microstructure, texture, and compositions, leading to, and even beyond, expected mechanical properties and functionalities. Although fabricating new materials and building up new techniques are of significant interest, it is also promising to utilize, quantify, and improve the excellent ability of SPD-induced structural and microstructural patterning through the use of existing SPD techniques. Such a material design approach not only creates new heterostructures improving the properties of the materials but also expands the hybrid and heterostructured UFG metals having SPD-induced patterning toward new applications of microfabrication which require nano-scale structural and functional components [106].

Figure 4.33 provides an excellent example of the potential use of SPD-induced patterning in microfabrication, where a schematic drawing of (a) a drone contains various sophisticated parts produced by microfabrication technologies, where the micro/nano-products include engines and reduction gear assemblies, sensors, and fasteners [107]. Figure 4.33b shows the typical microstructure of conventional engineering metals. When machined to have a small cogwheel, the grain size is not adequate for such a small part leading to a heterostructured component which is not applicable for the safety and reliability issues of the microsystem.

By contrast, SPD processing can produce reasonably homogeneous nanostructures which may be applied to form a UFG cogwheel, as shown in Fig. 4.33c.

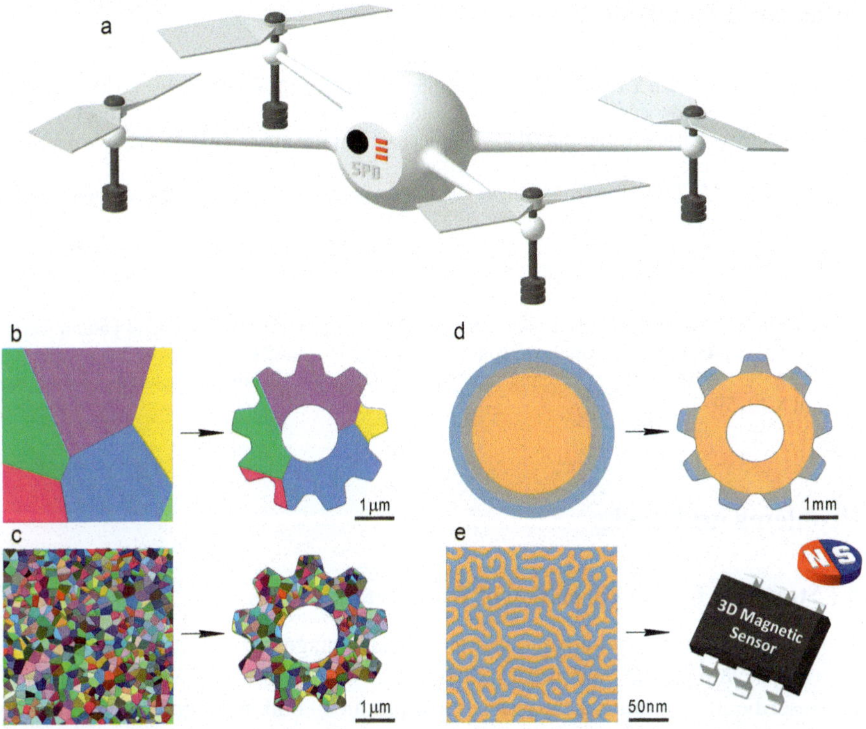

Fig. 4.33 Examples of the potential use of SPD-induced patterning in microfabrication. (Reprinted from [107], Copyright 2019, with permission from Elsevier)

Moreover, as shown in Fig. 4.33d, the formation of a bimetallic structure and hybrid structures (discussed in Sect. 4.1) can provide a smooth gradation in microstructure, and thus in properties and functionality, in the produced cogwheel. This can be especially beneficial for the possibility of mass production. Figure 4.33e shows a representative structure of a bulk nanocrystalline material mechanically alloyed by the SPD consolidation of powders, including magnetic materials and metal oxides. Optimized microstructures and compositions in the SPD-processed nanocrystalline hybrid metals can show excellent magnetic properties that may be applied for magnetoresistive sensors that are utilized in GPS navigation. Moreover, additional improved functionalities of electrical and optical properties, visible-light photocatalytic activities and dielectric constants (as discussed in Sect. 4.2) in nanostructured materials after SPD processing can be applied for a variety of micro-sensors and energy capacitors. The processing by SPD of bulk nanocrystalline materials, hybrid and heterostructured materials can therefore play an important role in interdisciplinary research fields such as Advanced Manufacturing, Applied Physics, Chemistry, Material Science, and Mechanical Engineering.

Suggested Reading

1. Alexandrov IV, Zhu YT, Lowe TC, Islamgaliev RK, Valiev RZ (1998) Consolidation of nanometer sized powders using severe plastic torsional straining. Nanostruct Mater 10:45–54
2. Meyers MA, Mishra A, Benson DJ (2006) Mechanical properties of nanocrystalline materials. Prog Mater Sci 51:427–556
3. Ma E, Zhu YT (2017) Towards strength–ductility synergy through the design of heterogeneous nanostructures in metals. Mater Today 20:323–331
4. Zhu YT, Ameyama K, Anderson PM, Beyerlein IJ, Gao H, Kim HS, Lavernia E, Mathaudhu S, Mughrabi H, Ritchie RO, Tsuji N, Zhang X, Wu X (2021) Heterostructured materials: superior properties from hetero-zone interaction. Mater Res Lett 9:1–31
5. Kulagin R, Beygelzimer Y, Bachmaier A, Pippan R, Estrin Y (2019) Benefits of pattern formation by severe plastic deformation. Appl Mater Today 15:236–241
6. Zhang X (2020) Heterostructures: new opportunities for functional materials. Mater Res Lett 8:49–59

Problems and Tasks

4.1 Strategies for the Formation of Bulk Metal Systems

1. List and describe the advantages and disadvantages of powder consolidations using severe plastic deformation techniques.
2. Describe the common differences in the produced bulk nanostructured materials through the powder metallurgy approach and from the bulk coarse metals.
3. Explain and analyze the advantages and technical difficulties of solid-state reaction for a set of dissimilar metals through the application of severe plastic deformation.
4. Explain the role of heating during SPD processing. Analyze the advantages and disadvantages of applying elevated temperatures during SPD of bulk metal systems.
5. From Sect. 4.1.2, list and describe the possible microstructural and structural changes during the solid-state reactions of dissimilar metals.

4.2. Functionalities of Hybrid Systems

6. Grain size is critical to understand the changes in mechanical properties. In case of nano-scale grains, often crystallite sizes measured by X-ray diffraction analysis are utilized instead of apparent grain size measured directly from micrographs. Explain the differences between crystallite size and grain size.
7. List and describe what mechanical properties are often measured in the bulk nanostructured materials. Also, describe the methods and techniques of measuring those properties with the difficulties and possible errors.
8. Explain the general anisotropic mechanical responses in the bulk nanostructured materials processed by (a) ARB, (b) ECAP, and (c) HPT.

4.3 Processing and Properties of Heterostructured Metals

9. Describe the difference between gradient microstructure and heterogeneous microstructure.
10. List and describe what microstructural and structural components can be heterogeneous in bulk nanostructured metals and alloys after SPD processing.
11. Heterogeneous nanostructures are often introduced by severe plastic deformation followed by short-term annealing. Describe the effect of the short-term heat treatment on the nanostructured materials.
12. Describe the fundamentals of architecturing of engineering metals and explain the difference from composite materials.

References

1. Koch CC (2003) Top-down synthesis of nanostructured materials: mechanical and thermal processing methods. Rev Adv Mater Sci 5:91–99
2. Korznikov AV, Safarov IM, Laptionok DV, Valiev RZ (1991) Structure and properties of superfine-grained iron compacted out of ultradisperse powder. Acta Metall Mater 39:3193–3197
3. Shen H, Li Z, Günther B, Korznikov AV, Valiev RZ (1995) Influence of powder consolidation methods on the structural and thermal properties of a nanophase Cu-50wt%Ag alloy. Nanostruct Mater 6:385–388
4. Alexandrov IV, Zhu YT, Lowe TC, Islamgaliev RK, Valiev RZ (1998) Microstructures and properties of nanocomposites obtained through SPTS consolidation of powders. Metall Mater Trans A 29:2253–2260
5. Alexandrov IV, Zhu YT, Lowe TC, Islamgaliev RK, Valiev RZ (1998) Consolidation of nanometer sized powders using severe plastic torsional straining. Nanostruct Mater 10:45–54
6. Alexandrov IV, Zhu YT, Lowe TC, Valiev RZ (1998) Severe plastic deformation: new technique for powder consolidation and grain size refinement. Powder Metall 41:11–13
7. Cubero-Sesin JM, Horita Z (2012) Powder consolidation of Al–10 wt% Fe alloy by high-pressure torsion. Mater Sci Eng A 558:462–471
8. Lee Z (2004) Microstructure and microhardness of cryomilled bulk nanocrystalline Al-7.5%Mg alloy consolidated by high pressure torsion. Scr Mater 51:209–214
9. Sort J, Zhilyaev A, Zielinska M, Nogues J, Surinach S, Thibault J et al (2003) Microstructural effects and large microhardness in cobalt processed by high pressure torsion consolidation of ball milled powders. Acta Mater 51:6385–6393
10. Valiev RZ, Mishral RS, Grozal J, Mukherjee AK (1996) Processing of nanostructured nickel by severe plastic deformation consolidation of ball-milled powder. Scr Mater 34:1443–1448
11. Stolyarov VV, Zhu YT, Lowe TC, Islamgaliev RK, Valiev RZ (2000) Processing nanocrystalline Ti and its nanocomposites from micrometer-sized Ti powder using high pressure torsion. Mat Sci Eng A Struct 282:78–85
12. Zhilyaev AP, Gimazov AA, Raab GI, Langdon TG (2008) Using high-pressure torsion for the cold-consolidation of copper chips produced by machining. Mat Sci Eng A Struct 486:123–126
13. Abd El Aal MI, Yoo Yoon E, Seop Kim H (2013) Recycling of AlSi8Cu3 alloy chips via high pressure torsion. Mater Sci Eng A 560:121–128
14. Yavari AR, Botta WJ, Rodrigues CAD, Cardoso C, Valiev RZ (2002) Nanostructured bulk Al90Fe5Nd5 prepared by cold consolidation of gas atomised powder using severe plastic deformation. Scr Mater 46:711–716

15. Botta WJ, Fogagnolo JB, Rodrigues CAD, Kiminami CS, Bolfarini C, Yavari AR (2004) Consolidation of partially amorphous aluminium-alloy powders by severe plastic deformation. Mat Sci Eng A Struct 375:936–941
16. Edalati K (2019) Review on recent advancements in severe plastic deformation of oxides by high-pressure torsion (HPT). Adv Eng Mater 21:1800272
17. Edalati K, Horita Z (2010) Application of high-pressure torsion for consolidation of ceramic powders. Scr Mater 63:174–177
18. Edalati K, Arimura M, Ikoma Y, Daio T, Miyata M, Smith DJ et al (2015) Plastic deformation of BaTiO3 ceramics by high-pressure torsion and changes in phase transformations, optical and dielectric properties. Mater Res Lett 3:216–221
19. Edalati K, Fujita I, Sauvage X, Arita M, Horita Z (2019) Microstructure and phase transformations of silica glass and vanadium oxide by severe plastic deformation via high-pressure torsion straining. J Alloy Compound 779:394–398
20. Tokunaga T, Kaneko K, Horita Z (2008) Production of aluminum-matrix carbon nanotube composite using high pressure torsion. Mat Sci Eng A Struct 490:300–304
21. Tokunaga T, Kaneko K, Sato K, Horita Z (2008) Microstructure and mechanical properties of aluminum-fullerene composite fabricated by high pressure torsion. Scr Mater 58:735–738
22. Menendez E, Sort J, Langlais V, Zhilyaev A, Munoz JS, Surinach S et al (2007) Cold compaction of metal-ceramic (ferromagnetic-antiferromagnetic) composites using high pressure torsion. J Alloy Compound 434:505–508
23. Menendez E, Salazar-Alvarez G, Zhilyaev AP, Surinach S, Baro MD, Nogues J et al (2008) Cold consolidation of metal-ceramic nanocomposite powders with large ceramic fractions. Adv Funct Mater 18:3293–3298
24. Matsuki K, Aida T, Takeuchi T, Kusui J, Yokoe K (2000) Microstructural characteristics and superplastic-like behavior in aluminum powder alloy consolidated by equal-channel angular pressing. Acta Mater 48:2625–2632
25. Senkov ON, Miracle DB, Scott JM, Senkova SV (2004) Equal channel angular extrusion compaction of semi-amorphous Al85Ni10Y2.5La2.5 alloy powder. J Alloy Compound 365:126–133
26. Robertson J, Im JT, Karaman I, Hartwig KT, Anderson IE (2003) Consolidation of amorphous copper based powder by equal channel angular extrusion. J Non-Cryst Solids 317:144–151
27. Quang P, Jeong YG, Yoon SC, Hong SH, Kim HS (2007) Consolidation of 1vol.% carbon nanotube reinforced metal matrix nanocomposites via equal channel angular pressing. J Mater Process Technol 187–188:318–320
28. Karaman I, Robertson J, Im JT, Mathaudhu SN, Luo ZP, Hartwig KT (2004) The effect of temperature and extrusion speed on the consolidation of zirconium-based metallic glass powder using equal-channel angular extrusion. Metall Mater Trans A 35a:247–256
29. Mathaudhu SN, Hartwig KT, Karaman I (2007) Consolidation of blended powders by severe plastic deformation to form amorphous metal matrix composites. J Non-Cryst Solids 353:185–193
30. Xia K, Wu X (2005) Back pressure equal channel angular consolidation of pure Al particles. Scr Mater 53:1225–1229
31. Jahedi M, Paydar MH (2010) Study on the feasibility of the torsion extrusion (TE) process as a severe plastic deformation method for consolidation of Al powder. Mat Sci Eng A Struct 527:5273–5279
32. Atkinson HV, Davies S (2000) Fundamental aspects of hot isostatic pressing: an overview. Metall Mater Trans A 31:2981–3000
33. Morris KJ (1991) Cold isostatic pressing. In: Concise encyclopedia of advanced ceramic materials. Pergamon Press, pp 84–88
34. Groza JR, Zavaliangos A (2000) Sintering activation by external electrical field. Mat Sci Eng A Struct 287:171–177
35. Bachmaier A, Hohenwarter A, Pippan R (2009) New procedure to generate stable nanocrystallites by severe plastic deformation. Scr Mater 61:1016–1019

36. Xia KN (2010) Consolidation of particles by severe plastic deformation: mechanism and applications in processing bulk ultrafine and nanostructured alloys and composites. Adv Eng Mater 12:724–729
37. Sauvage X, Jessner P, Vurpillot F, Pippan R (2008) Nanostructure and properties of a Cu-Cr composite processed by severe plastic deformation. Scr Mater 58:1125–1128
38. Sauvage X, Duchaussoy A, Zaher G (2019) Strain induced segregations in severely deformed materials. Mater Trans 60:1151–1158
39. Hajjari E, Divandari M, Razavi SH, Emami SM, Homma T, Kamado S (2011) Dissimilar joining of Al/Mg light metals by compound casting process. J Mater Sci 46:6491–6499
40. Sato YS, Park SHC, Michiuchi M, Kokawa H (2004) Constitutional liquation during dissimilar friction stir welding of Al and Mg alloys. Scr Mater 50:1233–1236
41. Beygelzimer Y, Estrin Y, Kulagin R (2015) Synthesis of hybrid materials by severe plastic deformation: a new paradigm of SPD processing. Adv Eng Mater 17:1853–1861
42. Sieber H, Wilde G, Perepezko JH (1999) Thermally activated amorphous phase formation in cold-rolled multilayers of Al-Ni, Al-Ta, Al-Fe and Zr-Cu. J Non-Cryst Solids 250:611–615
43. Eizadjou M, Talachi AK, Manesh HD, Shahabi HS, Janghorban K (2008) Investigation of structure and mechanical properties of multi-layered Al/Cu composite produced by accumulative roll bonding (ARB) process. Compos Sci Technol 68:2003–2009
44. Eivani AR, Taheri AK (2007) A new method for producing bimetallic rods. Mater Lett 61:4110–4113
45. Oh-ishi K, Edalati K, Kim HS, Hono K, Horita Z (2013) High-pressure torsion for enhanced atomic diffusion and promoting solid-state reactions in the aluminum-copper system. Acta Mater 61:3482–3489
46. Ahn B, Zhilyaev AP, Lee HJ, Kawasaki M, Langdon TG (2015) Rapid synthesis of an extra hard metal matrix nanocomposite at ambient temperature. Mat Sci Eng A Struct 635:109–117
47. Li Z, Zhang PF, Yuan H, Lin K, Liu Y, Yin DL et al (2016) Principle of one-step synthesis for multilayered structures using tube high-pressure shearing. Mat Sci Eng A Struct 658:367–375
48. Macwan A, Jiang XQ, Li C, Chen DL (2013) Effect of annealing on interface microstructures and tensile properties of rolled Al/Mg/Al tri-layer clad sheets. Mat Sci Eng A Struct 587:344–351
49. Lee KS, Lee YS, Kwon YN (2014) Influence of secondary warm rolling on the interface microstructure and mechanical properties of a roll-bonded three-ply Al/Mg/Al sheet. Mat Sci Eng A Struct 606:205–213
50. Beausir B, Scharnweber J, Jaschinski J, Brokmeier HG, Oertel CG, Skrotzki W (2010) Plastic anisotropy of ultrafine grained aluminium alloys produced by accumulative roll bonding. Mat Sci Eng A Struct 527:3271–3278
51. Mi YJ, Nie HH, Wang TL, Li XR, Hao XW, Liang W (2019) Effect of anisotropy on microstructures and mechanical properties of rolled Ti/Al/Mg/Al/Ti laminates. J Mater Eng Perform 28:4143–4151
52. Li L, Nagai K, Yin F (2008) Progress in cold roll bonding of metals. Sci Technol Adv Mater 9:023001
53. Mara NA, Beyerlein IJ (2015) Interface-dominant multilayers fabricated by severe plastic deformation: stability under extreme conditions. Curr Opin Solid State Mater Sci 19:265–276
54. Llorca-Isern N, Escobar AM, Roca A, Cabrera JM (2012) Equal channel angular pressing of Cu-Al bimetallic rod. Mater Sci Forum 706-709:1811–1816
55. Zhilyaev AP, Werner T, Cabrera JM (2020) Characterization of bimetallic interface in Cu-Al and Ni-Cu rods cold welded by equal channel angular pressing. Adv Eng Mater 22. https://doi.org/10.1002/adem.201900653
56. Buet E, Dubois J-B, Olier P, Thilly L, Lecouturier F, Renault P-O (2011) Microstructure and texture of copper/niobium composites processed by ECAE. Int J Mater Form 5:121–127
57. Qi YS, Lapovok R, Estrin Y (2016) Microstructure and electrical conductivity of aluminium/steel bimetallic rods processed by severe plastic deformation. J Mater Sci 51:6860–6875
58. Bouaziz O, Kim HS, Estrin Y (2013) Architecturing of metal-based composites with concurrent nanostructuring: a new paradigm of materials design. Adv Eng Mater 15:336–340

59. Han JK, Liss KD, Langdon TG, Kawasaki M (2019) Synthesis of a bulk nanostructured metastable Al alloy with extreme supersaturation of Mg. Sci Rep 9:17186
60. Edalati K (2019) Metallurgical alchemy by ultra-severe plastic deformation via high-pressure torsion process. Mater Trans 60:1221–1229
61. Han JK, Herndon T, Jang JI, Langdon TG, Kawasaki M (2020) Synthesis of hybrid nanocrystalline alloys by mechanical bonding through high-pressure torsion. Adv Eng Mater 22:201901289
62. Miyazaki T, Terada D, Miyajima Y, Suryanarayana C, Murao R, Yokoyama Y et al (2011) Synthesis of non-equilibrium phases in immiscible metals mechanically mixed by high pressure torsion. J Mater Sci 46:4296–4301
63. Rigney DA, Karthikeyan S (2010) The evolution of tribomaterial during sliding: a brief introduction. Tribol Lett 39:3–7
64. Kulagin R, Beygelzimer Y, Iyanisenko Y, Mazilkin A, Straumal B, Hahn H (2018) Instabilities of interfaces between dissimilar metals induced by high pressure torsion. Mater Lett 222:172–175
65. Wang JT, Li Z, Wang J, Langdon TG (2012) Principles of severe plastic deformation using tube high-pressure shearing. Scr Mater 67:810–813
66. Edalati K, Horita Z, Fujiwara H, Ameyama K (2010) Cold consolidation of ball-milled titanium powders using high-pressure torsion. Metall Mater Trans A 41a:3308–3317
67. Edalati K, Toh S, Watanabe M, Horita Z (2012) In situ production of bulk intermetallic-based nanocomposites and nanostructured intermetallics by high-pressure torsion. Scr Mater 66:386–389
68. Edalati K, Toh S, Iwaoka H, Watanabe M, Horita Z, Kashioka D et al (2012) Ultrahigh strength and high plasticity in TiAl intermetallics with bimodal grain structure and nanotwins. Scr Mater 67:814–817
69. Ganduglia-Pirovano MV, Hofmann A, Sauer J (2007) Oxygen vacancies in transition metal and rare earth oxides: current state of understanding and remaining challenges. Surf Sci Rep 62:219–270
70. Razavi-Khosroshahi H, Edalati K, Arita M, Horita Z, Fuji M (2016) Plastic strain and grain size effect on high-pressure phase transformations in nanostructured TiO2 ceramics. Scr Mater 124:59–62
71. Razavi-Khosroshahi H, Edalati K, Wu J, Nakashima Y, Arita M, Ikoma Y et al (2017) High-pressure zinc oxide phase as visible-light-active photocatalyst with narrow band gap. J Mater Chem A 5:20298–20303
72. Pankove JI (1971) Optical processes in semiconductors. Dover Publications
73. Beyerlein IJ, Mara NA, Carpenter JS, Nizolek T, Mook WM, Wynn TA et al (2013) Interface-driven microstructure development and ultra high strength of bulk nanostructured Cu-Nb multilayers fabricated by severe plastic deformation. J Mater Res 28:1799–1812
74. Kawasaki M, Han JK, Lee DH, Jang J, Langdon TG (2018) Fabrication of nanocomposites through diffusion bonding under high-pressure torsion. J Mater Res 33:2700–2710
75. Luo D, Huminiuc T, Huang Y, Polcar T, Langdon TG (2020) The fabrication of high strength Zr/Nb nanocomposites using high-pressure torsion. Mat Sci Eng A Struct 790:139693
76. Ahn B, Lee HJ, Choi IC, Kawasaki M, Jang JI, Langdon TG (2016) Micro-mechanical behavior of an exceptionally strong metal matrix nanocomposite processed by high-pressure torsion. Adv Eng Mater 18:1001–1008
77. Bazarnik P, Bartkowska A, Romelczyk-Baishya B, Adamczyk-Cieslak B, Dai JY, Huang Y et al (2020) Superior strength of tri-layered Al-Cu-Al nano-composites processed by high-pressure torsion. J Alloy Compound 846:156380
78. Hernández-Escobar D, Rahman ZU, Yilmazer H, Kawasaki M, Boehlert CJ (2018) Microstructural evolution and intermetallic formation in Zn-Mg hybrids processed by high-pressure torsion. Philos Mag 99:557–584
79. Han JK, Han DK, Liang GY, Jang JI, Langdon TG, Kawasaki M (2018) Direct bonding of aluminum-copper metals through high-pressure torsion processing. Adv Eng Mater 20:1800642

80. Meyers MA, Mishra A, Benson DJ (2006) Mechanical properties of nanocrystalline materials. Prog Mater Sci 51:427–556
81. Meyers MA, Chen P-Y (2014) Biological materials science: biological materials, bioinspired materials, and biomaterials. Cambridge University Press
82. Liddicoat PV, Liao XZ, Zhao Y, Zhu Y, Murashkin MY, Lavernia EJ et al (2010) Nanostructural hierarchy increases the strength of aluminium alloys. Nat Commun 1:63
83. Lu K (2014) Nanomaterials. Making strong nanomaterials ductile with gradients. Science 345:1455–1456
84. Fang TH, Li WL, Tao NR, Lu K (2011) Revealing extraordinary intrinsic tensile plasticity in gradient nano-grained copper. Science 331:1587–1590
85. Wu X, Jiang P, Chen L, Yuan F, Zhu YT (2014) Extraordinary strain hardening by gradient structure. Proc Natl Acad Sci U S A 111:7197–7201
86. Long JZ, Pan QS, Tao NR, Dao M, Suresh S, Lu L (2019) Improved fatigue resistance of gradient nanograined Cu. Acta Mater 166:56–66
87. Ma E, Zhu T (2017) Towards strength-ductility synergy through the design of heterogeneous nanostructures in metals. Mater Today 20:323–331
88. Valiev RZ, Alexandrov IV, Zhu YT, Lowe TC (2002) Paradox of strength and ductility in metals processed by severe plastic deformation. J Mater Res 17:5–8
89. Lu K, Lu J (1999) Surface nanocrystallization (SNC) of metallic materials-presentation of the concept behind a new approach. J Mater Sci Technol 15:193–197
90. Li WL, Tao NR, Lu K (2008) Fabrication of a gradient nano-micro-structured surface layer on bulk copper by means of a surface mechanical grinding treatment. Scr Mater 59:546–549
91. Jing LJ, Pan QS, Long JZ, Tao NR, Lu L (2019) Effect of volume fraction of gradient nanograined layer on high-cycle fatigue behavior of Cu. Scr Mater 161:74–77
92. Liu Y, Jin B, Lu J (2015) Mechanical properties and thermal stability of nanocrystallized pure aluminum produced by surface mechanical attrition treatment. Mat Sci Eng A Struct 636:446–451
93. Fang XT, He GZ, Zheng C, Ma XL, Kaoumi D, Li YS et al (2020) Effect of heterostructure and hetero-deformation induced hardening on the strength and ductility of brass. Acta Mater 186:644–655
94. Wu XL, Zhu YT (2017) Heterogeneous materials: a new class of materials with unprecedented mechanical properties. Mater Res Lett 5:527–532
95. Fu ZQ, MacDonald BE, Li ZM, Jiang ZF, Chen WP, Zhou YZ et al (2018) Engineering heterostructured grains to enhance strength in a single-phase high-entropy alloy with maintained ductility. Mater Res Lett 6:634–640
96. Wu X, Yang M, Yuan F, Wu G, Wei Y, Huang X et al (2015) Heterogeneous lamella structure unites ultrafine-grain strength with coarse-grain ductility. Proc Natl Acad Sci U S A 112:14501–14505
97. Li JS, Cao Y, Gao B, Li YS, Zhu YT (2018) Superior strength and ductility of 316L stainless steel with heterogeneous lamella structure. J Mater Sci 53:10442–10456
98. Wang YF, Huang CX, He Q, Guo FJ, Wang MS, Song LY et al (2019) Heterostructure induced dispersive shear bands in heterostructured Cu. Scr Mater 170:76–80
99. Liu YF, Cao Y, Mao QZ, Zhou H, Zhao YH, Jiang W et al (2020) Critical microstructures and defects in heterostructured materials and their effects on mechanical properties. Acta Mater 189:129–144
100. Wang YF, Huang CX, Wang MS, Li YS, Zhu YT (2018) Quantifying the synergetic strengthening in gradient material. Scr Mater 150:22–25
101. Gao B, Lai Q, Cao Y, Hu R, Xiao L, Pan Z et al (2020) Ultrastrong low-carbon nanosteel produced by heterostructure and interstitial mediated warm rolling. Sci Adv 6:eaba8169
102. Su J, Raabe D, Li Z (2018) Hierarchical microstructure design to tune the mechanical behavior of an interstitial TRIP-TWIP high-entropy alloy. Acta Materialia. https://doi.org/10.1016/j.actamat.2018.10.017
103. Zhang X (2019) Heterostructures: new opportunities for functional materials. Mater Res Lett 8:49–59

104. Kang JY, Kim JG, Park HW, Kim HS (2016) Multiscale architectured materials with composition and grain size gradients manufactured using high-pressure torsion. Sci Rep 6:26590
105. Beygelzimer Y, Kulagin R, Estrin Y (2019) Severe plastic deformation as a way to produce architectured materials. In: Architectured materials in nature and engineering. Springer, pp 231–255
106. Xu J, Wang XW, Wang CJ, Yuan L, Chen WJ, Bao JX et al (2021) A review on micro/nanoforming to fabricate 3D metallic structures. Adv Mater 33:e2000893
107. Kulagin R, Beygelzimer Y, Bachmaier A, Pippan R, Estrin Y (2019) Benefits of pattern formation by severe plastic deformation. Appl Mater Today 15:236–241

Chapter 5
Innovations and Commercialization of UFG Materials

The application and commercialization of UFG materials are associated with three important advantages: the superior properties of these nanomaterials, their efficient fabrication, and the possibility of producing cutting-edge products from these materials. Over the last decade, active development has been underway for fabricating pilot articles from different nanometals, the shaping of which included successful realization of mechanical processing, electrochemical treatment, and superplastic forming [1–3]. Recent reports documented several market areas for nanostructured metals [2–4], and it is evident that many of these new structural and functional applications involve extreme environments where exceptional strength and other specific properties are needed. In this chapter, we consider several examples of UFG materials' development for their innovation applications in engineering and medicine.

5.1 Nanometals for Medical Devices

Presently, over 70% of implant devices are constructed from metallic materials because of their strength, toughness, and durability [5]. On the other hand, another principal concern in the use of metals in the human body is their safety in terms of toxicity with metal ion release, which in turn justifies the application of corrosion-resistance materials, mostly such as stainless steel, CoCrMo alloy, CP Ti and Ti alloys, and more recently—bioresorbable Mg and Fe alloys.

New alloys with optimized chemical composition, manufacturing processes, and surface modifications undergo continuous research and development to satisfy the clinical demands for medical devices [6, 7]. At the same time, during the last decade, the nanostructuring of metals to improve their properties using severe plastic deformation (SPD) processing [8, 9] has become a new and promising area of modern biomaterials science and engineering.

R. Z. Valiev et al., *Ultrafine-Grained Materials*, The Minerals, Metals & Materials Series, https://doi.org/10.1007/978-3-031-31729-3_5

As demonstrated in Sect. 5.3, from the perspective of medical applications, the use of SPD processing techniques for the formation of nanostructures in metals and alloys may positively affect both mechanical and biomedical properties. For the latter, surface modification of bulk nanomaterials also plays an important role. All of this provides the possibility for the development of medical devices with improved design and functionality.

Let us consider two examples of making and using these implants from nanoTi with premium design for dentistry and orthopedics [10].

5.1.1 Nano Ti Miniplates

Change of materials, such as the case with substituting nanostructured CP Ti for coarse-grained (CG) Ti, leads to altering the design of devices and this is when a set of simple rules should be taken into account. Lately, the necessary calculations were conducted to analyze the geometry of miniplates made of nanostructured Ti for maxillofacial surgery [10, 11].

A mini-plate made of CP Ti, as specified by ASTM F 67, was used by a "Conmet" company (Moscow, Russia) as the starting point for redesigning the product dimensions for making a nanostructured CP Ti mini-plate (Fig. 5.1) [11]. The cross-section properties of a new plate were computed using the estimates of the fatigue endurance limit for coarse-grained Grade 4 Ti $\sigma_{f(CG)}$ and nanostructured Grade 4 Ti $\sigma_{f(NS)}$. A comparison of bending strength [12, 13] for mini-plates from conventional CG Ti and nanostructured Ti was performed because plates commonly support bending loads [14]. Further, the fatigue strength of mini-plates was compared through testing with an ElectroPuls E3000 system [13]. The number of cycles to failure N was determined through cyclic 3 mm displacements at 30 Hz. Triplicate tests were completed.

Table 5.1 provides the estimation results for the cross-section area of the nano-Ti plate [11]. The width and the diameter remain the same as in the standard item, whereas the plate thickness alters from 0.9 to 0.7. These experiments showed that the plate strength for nano Ti does not necessarily decrease in the process of bending even when the cross-section area is reduced. Bending tests were completed for the experimental values of the fatigue life of plates, as is shown in Fig. 5.2 [11]. The standard plates have sustained 17,000 ± 500 cycles and the redesigned plate made from nanostructured Ti could survive a larger number of cycles to failure

Fig. 5.1 Image of a mini plate with six holes made of nanostructured Grade 4 Ti. (Reproduced from [11], Copyright 2016, with permission of John Wiley & Sons)

Table 5.1 Sizes of a base plate from CG and NS CP titanium Reproduced from [11], Copyright 2016, with permission of John Wiley & Sons

		Strength ratio, K_r	Geometrical sizes of a plate				Axial resistance moment (W)
Plate	Fatigue strength limit σ_f, MPa	$K_r = \sigma_{f(CG)}/\sigma_{f(NS)}$	Length, L, mm	Area in the central part, A, mm^2	Thickness, h, mm	Width, b (mm)	$W = \frac{bh^2}{6} mm^3$
Ti plate	340	0.6	46.4	2.7	0.9	3	0.40
Nano-Ti plate	620		46.4	2.1	0.7	3	0.25

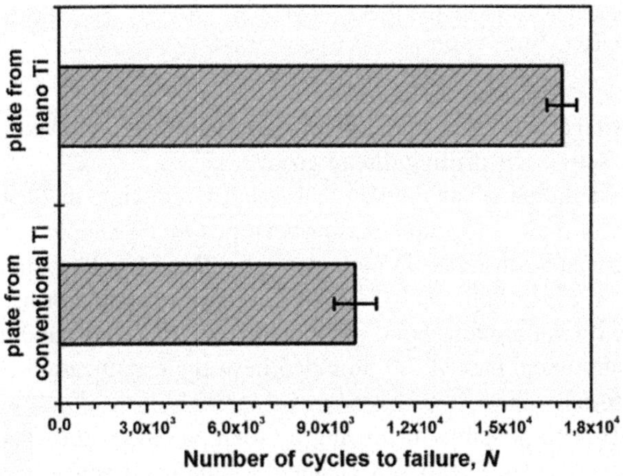

Fig. 5.2 Fatigue life of the standard plate and redesigned plate from nano-Ti. (Reproduced from [11], Copyright 2016, with permission of John Wiley & Sons)

(105,000 ± 800). This result indicates that the plate from nano-Grade 4 Ti has its bending strength significantly improved and thus, speaks of its clear advantage over the standard item manufactured from coarse-grained Ti.

5.1.2 *Dental Thin Implants*

Numerous nanoimplant® dental implants have been produced from pure nanostructured Grade 4 Ti rods processed by the ECAP-Conform on CNC machines in diameters of 2.0 mm, 2.4 mm, and 3.5 mm by the company "Timplant" s.r.o., Czech Republic over the last years (http://www.timplant.cz/en/) (see also Ref. [10]). The implants are characterized by an etched threaded tapered intraosseal part with a polished gingival part and prosthetics cone top with interior thread above it, and

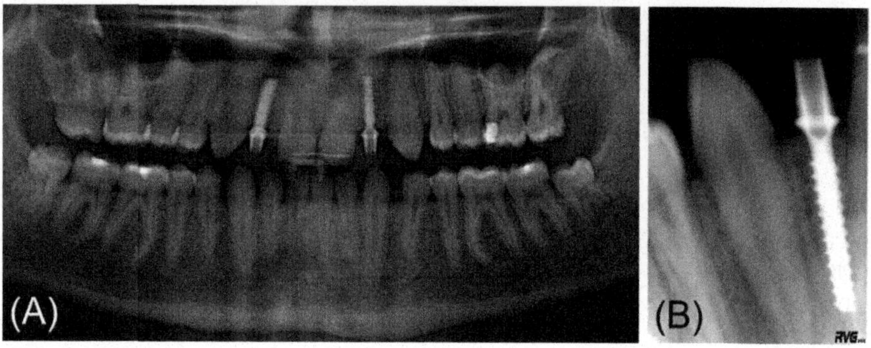

Fig. 5.3 2.0 mm diameter nanoimplant (www.timplant.cz/en/) from nanostructured Grade 4 Ti in a panoramic X-ray radiograph after surgery (**a**), the right one, and the control radiograph obtained after incorporation of the implant (**b**). (Reproduced from [15], Copyright 2015, with permission of John Wiley & Sons)

with anti-rotation crown element. High primary stability is achieved by an enhanced thread design with a self-drilling thread groove.

The reduced radius of an implant allows for reducing medical interference, therefore trauma related to implantation becomes much smaller. Implants with a reduced radius can be installed in patients with a thin alveolar bone, while in such cases, the use of a conventional CG Ti implant is not possible or necessitates extra intervention with subsequent bone augmentation [10]. Thin nanoimplants with a diameter of intraosseal part of 2.0 mm also have their application in dental orthodontics (Fig. 5.3).

Nano-CP Ti is responsible for having no toxic alloying elements (such as V or Al) or allergens (such as Ni, Co, or Cr), and it is clearly attractive for the production of thin implants owing to its high strength. Statistical observation completed over 2 years by five surgeon dentists both from state-owned and privately run dental surgeries in the Czech Republic [15] demonstrated high clinical performance of 2.4 mm Nanoimplants®.

Surface properties are also critical for medical implants and their modification has proved to accelerate osseointegration. In particular, pure Ti is bionert material, meaning that when placed in the human body, it has minimal interaction with surrounding tissue, including the human bone. Application of SPD processing to achieve nano-sized grains in CP Ti allows various types of bone-forming cells to adhere and proliferate with better efficiency and this was demonstrated in numerous studies [16, 17]. Additional surface modification may further contribute to the better bioactive performance of implants made of nano Ti. Recently, two main approaches have been proposed for dental implant surface modification: chemical etching and deposition of bioactive coatings [10, 18–21].

The topography of etched surfaces is strongly determined by etching solution and etching time. Different solutions, such as acidic (H_2SO_4/H_2O_2) or basic (NH_4OH/H_2O_2) Piranha solutions can be used for the etching of CP Ti [22], resulting in different surface topographies. The surface topography was found to be clearly modified after varying etching time, and the effect was more pronounced in the nano Ti.

Biocompatible coatings can also efficiently make the nano Ti implants integrate with the human bone [10]. Presently, the research into the synthesis of biocompatible coatings integrating the inorganic (Ca-, P- containing phases) and organic (biologically active and bioinert molecules) components on titanium implants appears to be a topical state of the art [21] (see also Sect. 3.2).

The results of recent studies convincingly demonstrate the fact that metallic biomaterials (Ti and Ti alloys, stainless steels, and others), after their nanostructuring by severe plastic deformation techniques, demonstrate improved mechanical and functional properties, and this accounts for the considerable interest in their medical applications [23]. Improved mechanical properties of nanomaterials create solid opportunities for manufacturing medical implants with improved design and miniaturized devices. The examples of the first successful applications of dental and maxillofacial nanoimplants made of nanostructured titanium are considered and discussed above.

The study of biomedical properties of SPD-processed nanometals is a relatively new task because nanostructuring introduces certain advanced features in the materials biological response. At the same time, the results of recent studies on surface modification, including chemical etching and deposition of bioactive coatings, make it a promising approach for the development of advanced medical implants and devices with improved design and biofunctionality.

5.2 Nanostructured Cu and Al Alloys for Advanced Electrical Conductors

Copper and its alloys, as well as aluminum and its alloys, are widely used in electrical, transport, and nuclear industries. The most important requirement imposed on these alloys is a combination of high strength, and electrical and thermal conductivity. Figure 5.4 shows examples of properties typical of modern Cu-based materials [24, 25].

In Sect. 3.1, it was demonstrated that nanostructuring of Cu, Al, and their alloys can significantly increase their mechanical properties with no reduction in their electrical conductivity or even with its further improvement. The development of metallic conductors with such attractive electrical conductivity is one of the most topical subjects in modern electrical engineering.

In this regard, the problem of controlling the contribution of various structural parameters affecting the strength, electrical and thermal conductivity, in order to achieve the optimum of the indicated properties of UFG copper and aluminum alloys, is relevant in terms of the prospects of application. Another topical problem is the development of SPD techniques for the formation of a set of the mentioned properties under industrial production conditions.

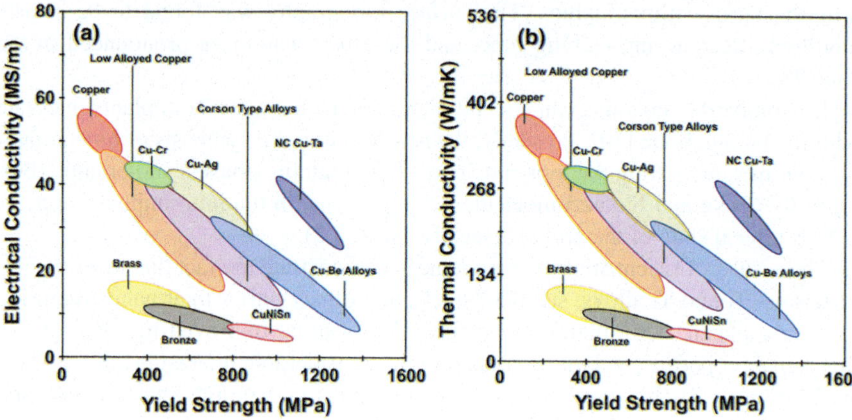

Fig. 5.4 (**a**) Electrical and (**b**) thermal conductivities at room temperature versus 0.2% yield strength of pure Cu and various Cu-based alloys. The variation in the NC Cu-Ta strength and conductivities are due to different Ta concentrations (1 at%, 5 at%, and 10 at% as well as changes in the average grain size). Note: literature electrical conductivity data were obtained from [24]. (Reprinted from [24], Copyright 2019, with permission from Elsevier)

Let's consider several promising directions for the application of products from these UFG materials.

Copper electrodes are widely used in welding copper and copper alloys as well as cast iron. The main requirements for such electrodes are high electrical and thermal conductivity. Low strength makes copper electrodes made from CG copper and copper alloys tend to have high wear. However, electrodes made of UFG copper [26] by SPD techniques demonstrate increased strength, sufficiently high electric and thermal conductivity, reduced and more uniform wear of the electrode surface.

Improved mechanical properties and electrical conductivity can be achieved by creating nanocomposites consisting of Cu with dispersed WC nanoparticles [27]. The nanoparticles can effectively induce hardening, while having less negative effect on electrical and thermal conductivity than the solute atoms. Cu-WC nanocomposites exhibit high levels of hardness and strength, Young's modulus, and thermal stability, while maintaining high levels of electrical and thermal conductivity. For example, high levels of tensile strength (over 600 MPa) and electrical conductivity (over 85% IACS) were simultaneously achieved [27]. Cu-WC nanocomposites can be used to manufacture products such as current carrying cores, electric motors, contact wires for high-speed railroads, heat exchangers, and electrodes for resistance welding.

The SPD technique of accumulated roll bonding has great prospects of application for series production in industrial conditions of high-strength electrically conductive sheets made of multilayer Cu/Nb composites (Fig. 5.5) [28]. This technique makes it possible to obtain continuous homogeneous lamellar structures with alternating Cu and Nb layers of adjustable thickness from micrometers to several tens of nanometers (Fig. 5.6).

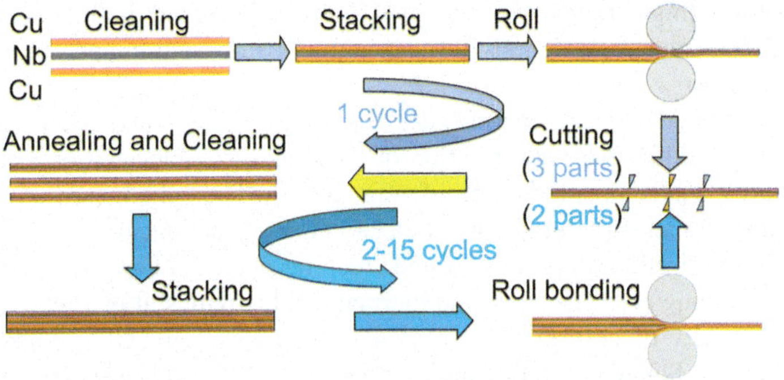

Fig. 5.5 Schematic illustration for the fabrication of Cu/Nb multilayer composites by modified ARB process. (Reprinted from [28], Copyright 2021, with permission from Elsevier)

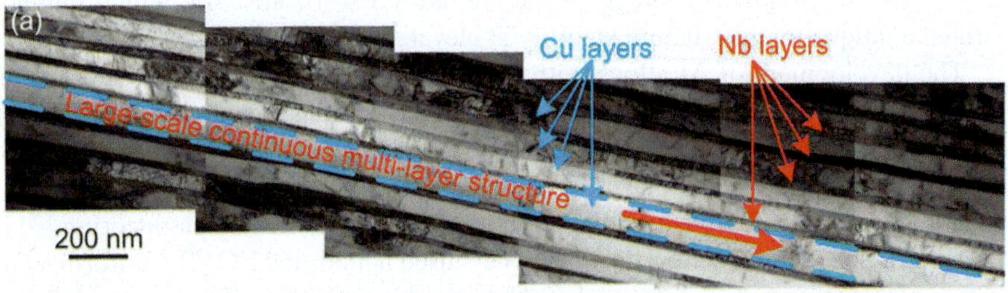

Fig. 5.6 (**a**) Cross-sectional TEM images of the Cu/Nb nanolaminate composites after processing by ARB through 13 cycles. (Reprinted from [28], Copyright 2021, with permission from Elsevier)

The composite is characterized by enhanced mechanical properties (ultimate tensile strength of 1.2 GPa at a thickness of about 16 nm) and high electrical conductivity (over 60% IACS), virtually independent of layer thickness. The high strength is due to alternating layers with a thickness of several tens of nanometers, and regular accumulations of dislocations on the surfaces between the layers. The low extent of grain boundaries and interfaces in continuous Cu layers, absence of dislocations and dispersed separation are responsible for the high values of electrical conductivity.

Creating such structures is cost-effective and scalable under industrial conditions. This approach can easily be applied to produce other advanced structural composites and multifunctional materials for heat transfer as well as catalytic and radiation-resistant applications.

Annealed copper typically exhibits a low electrical resistivity of $1.72 \cdot 10^{8}$ Ω·m (100% IACS). However, the strength and thermal stability of annealed copper and related alloys are relatively low. This limits the use of copper as a material for high-speed pulse magnets [29] or heat exchangers, where the heat resistance of metal-based alloys is generally extremely low, i.e., on the order of 30–50 MPa at 800 °C for the best nickel-based heat exchanger currently in use [30]. New alloys for next-generation heat exchangers, high-tech electronics, high-speed contact wires for rails and high-speed pulse magnets should combine extremely high strength, low strain rate sensitivity (high creep resistance), and high electrical and thermal conductivity. Such materials must be able to withstand high pressures and temperatures for several tens of hours.

Nanostructured copper electrical alloys have high strength and electrical conductivity (Chap. 3); however, deformation of these alloys at elevated or intermediate temperatures is accompanied by rapid grain growth. Thus, the increase in the thermal stability of nanostructured copper electrical alloys is topical.

To retain high strength properties and creep resistance at elevated temperatures, while maintaining high electrical and thermal conductivities, it is important to suppress the activity of deformation processes characteristic of UFG copper alloys, which are grain boundary sliding and grain rotation [24]. The suppression of these processes was realized by forming clusters of alloying elements as a result of short-circuit solute diffusion along the grain boundaries. The formed clusters allowed to prevent the GB migration and to stabilize the nanostructured state formed by SPD. This development resulted in the NC alloy Cu–10 at.% Ta, which demonstrated a unique microstructure stability at elevated temperatures.

The development of Al alloys with UFG structure is of great interest for overhead power lines (see also Chap. 3). Russia's leading industrial aluminum company RUSAL showed interest in the work of Russian scientists from Ufa back in 2014. These teams jointly developed a technology for producing nanostructured aluminum alloys with increased structural strength for use in electrical engineering [31].

To produce UFG Al wires, the researchers used a modified ECAP-Conform technique. Conform is a processing technique to produce wire by subjecting the billet to continuous extrusion [https://engitime.ru/tehnologi/texnologiya-conform.html]. The combined technique is called ECAP-Conform or ECAP-C (Fig. 5.7) [32]. When implementing the ECAP-C technique, the billet is pushed through the die set by the friction force, as well as during the traditional Conform process. This process can be implemented many times. In this process, the workpiece is driven forward to

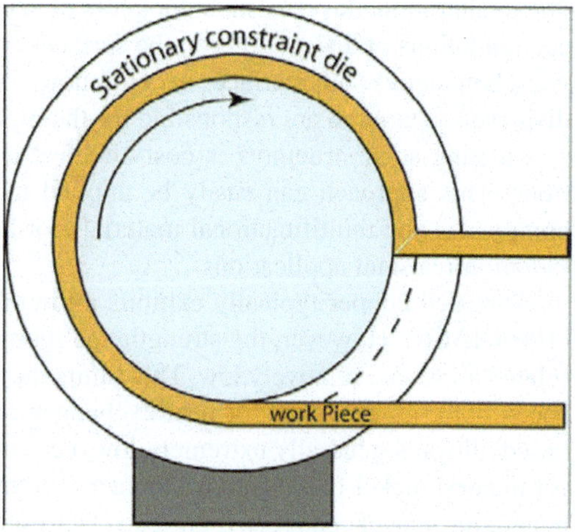

Fig. 5.7 The principle of ECAP-C technique. (Reprinted from [32], Copyright 2004, with permission from Elsevier)

a groove made on the rotating shaft by the friction force and is directed to the die set for the ECAP.

The long-length billets of Al alloy 6101 obtained by ECAP-C technique and further drawing had a diameter of 3.2 mm. The microstructure of the alloy was characterized by an average grain size of less than 500 nm and precipitated nano-sized particles of the secondary phase Mg_2Si. The ultimate tensile strength was 320–370 MPa, elongation to failure was not less than 3%, and specific electrical resistance amounted to 0.0302–0.0307 Ohm·mm^2/m, which corresponds to 57.2–56.1% IACS.

Physical and mechanical properties demonstrated by pilot samples obtained from wire subjected to ECAP-C were much higher in comparison with all types of wire (Al2–Al7) produced industrially in accordance with the international standard [BS EN 50183:2000, "Conductors for overhead lines – aluminum-magnesium-silicon alloy wires," published standards, May 2000].

The presented results as well as a number of other examples (see, for example, [3, 24] and others) show that SPD processing, and the ECAP-Conform in particular, can be successfully used for Cu and Al alloys in order to produce electrical conductors in the form of bars and wires with superior mechanical and physical properties.

5.3 Ultrafine-Grained Metallic Alloys for Hydrogen Storage

According to some estimates, by 2050, 18% of the world's energy needs will be met by hydrogen technology. Particular attention to hydrogen energy is caused by nearly unlimited resources of hydrogen, the absence of harmful emissions into the atmosphere, the high efficiency of hydrogen engines, etc. [33]. At the same time, the most important topical problems of hydrogen power engineering are the production, storage, and transportation of hydrogen. Presently, the latter is carried out mainly in gaseous or liquid states. However, hydrogen gas has an adverse effect on the structural properties of the materials used to store and transport gaseous hydrogen. At the same time, hydrogen, which actively interacts with oxygen, poses a great danger.

There are high expectations for an alternative method of hydrogen storage in the solid state. This method has advantages over the above-mentioned ones. It also addresses the issue of the safety of hydrogen engines for cars, airplanes, rockets, etc.

To store hydrogen in a solid state, it is necessary to use materials that easily bind hydrogen, have high hydrogen storage capacity, and easily release hydrogen when heated. In this case, the interaction of hydrogen with such substances should lead to the formation of light metal hydrides or intermetallic compounds. One of the few compounds that meet the standards stipulated by the largest hydrogen programs for renewable energy storage systems is magnesium hydride MgH_2 and some Mg alloys. This is due to their capability to absorb hydrogen up to 7.6 wt.%, low specific gravity, and low cost. However, the high thermodynamic stability ($\Delta H = -75$ kJ/mol), high hydrogen desorption temperature (above 400 °C), and relatively weak kinetics of hydrogen absorption and desorption at temperatures below 350 °C prevent the industrial application of Mg alloys. The low hydrogen absorption/desorption rates pose a particular problem.

The solution to the problem of hydrogen storage in the solid state can be found by creating a high extent of grain boundaries and the density of crystalline structure defects by SPD techniques and the formation of UFG structures in Mg and its alloys [33, 34].

This approach was initiated by revealing the increased ability to accumulate hydrogen as a result of high-energy mechanical alloying, leading to the formation of Mg powders [35]. Reduction of the grain size to nanocrystalline (10–30 nm) leads to the significant acceleration of the kinetics of hydrogen absorption. The indicated growth is caused by an increase in the volume fraction of the grain boundaries. At the same time, the hydrogen desorption temperature decreases by almost 100 °C.

At the same time, the use of powders obtained by high-energy ball milling (HEBM) for hydrogen mass storage is characterized by a number of drawbacks and limitations, including the high cost of the technique, low productivity, energy consumption, harmful environmental impact, and thus the implementation of extensive precautions to minimize environmental impact and health risks. Mg and Mg alloy powders subjected to ball milling rarely meet the requirements of stability with respect to the absorbed hydrogen. The kinetics of hydrogen absorption/desorption processes with a large number of cycles is also unstable, at least without catalysts. However, the need to minimize the contamination of powders with oxygen and other impurities during refinement is the main problem.

The increase of hydrogen absorptivity, better absorption and desorption kinetics, and storage stability is achieved by using different SPD techniques such as ECAP, HPT, etc., as applied to bulk samples of Mg alloys.

It is demonstrated [36] that $Mg_{70}Ni_{30}$ powders subjected to refinement in a ball mill and disks condensed by HPT of refined powders absorb hydrogen almost without activation. At the same time, the maximum capacity is reached after the first two absorption and desorption cycles. The maximum hydrogen content in the powders is about 2 and 2.4 wt.% after refinement in a ball mill for 1 and 10 h, respectively (Fig. 5.8). The higher hydrogen content corresponds to a longer ball milling due to the formation of a finer structure at both micron and nanoscale levels. The hydrogen desorption rate in powders subjected to high-energy ball milling increases by several orders of magnitude compared to the state before milling.

HPT leads to noticeably increased hydrogen absorption. The maximum hydrogen content is 3 wt.%, which corresponds to the formation of the compound $Mg_2NiH_{3.3}$ and almost complete hydrogenation. The increased hydrogen absorption, on the one hand, is explained by the smaller grain size and, consequently, by the increased specific surface area of GBs. On the other hand, the HPT leads to the formation of numerous crystal structure defects (e.g., dislocations) near which the lattice density is reduced and which can absorb hydrogen adjusted for diffusion processes. Numerous crystal structure defects in consolidated disks significantly contribute to locations for hydrogen.

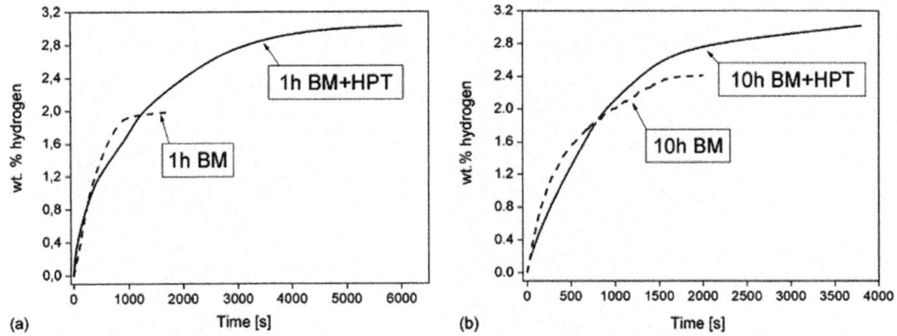

Fig. 5.8 Hydrogen kinetic absorption curves of Mg70Ni30 powder milled for 1 h and then processed by HPT at 573 K (**a**) and powder milled for 10 h and then deformed by HPT at 573 (**b**). (Reprinted from [36], Copyright 2010, with permission from Elsevier)

The increased hydrogen absorptivity was also found due to HPT of $MgNi_2$ alloy, as compared to its initial CG state [34].

The possibility of obtaining bulk UFG billets of Mg and its alloys using such SPD techniques as ECAP and its varieties is of particular interest from the perspective of creating effective hydrogen storage tanks.

ECAP of massive billets leads to the formation of UFG structures in Mg and its alloys. These microstructures have the above prerequisites for hydrogen capacities. They are characterized by grain sizes of tens to hundreds of nanometers as well as a high extent of nonequilibrium GBs and high densities of crystalline structure defects.

Figure 5.9 shows bright-field TEM micrographs of an as-cast $Mg_{89}Ni_{11}$ sample (a, b), a sample processed by a single ECAP pass (c, d), and one processed by ten ECAP passes (e, f) [37]. The dashed lines in Fig. 5.9 indicate the individual Mg grains and Mg_2Ni particles. As can be seen, ten ECAP passes along the route B_C with a decrease in temperature from 300 to 230 °C leading to refinement of the initial as-cast structure to nanometer size (Fig. 5.9).

Microstructure refinement in $Mg_{89}Ni_{11}$ alloy by ECAP leads to an increase in the kinetics of hydrogen absorption/desorption.

ECAP led to an impressive increase in gravitometric capacity and kinetics of hydrogen desorption in bulk billets of ZK60B alloy (Fig. 5.10) [38]. This effect is brightly demonstrated by the red curve corresponding to the state with the minimum of the considered grain size. In this case, the gravitometric hydrogen capacity exceeds 6.5 wt.%.

The rate of hydrogen desorption in this case strongly depends on the grain size. At the same time, the addition of a small amount of Cr as a catalyst to the alloy at the casting stage contributes to the acceleration of hydrogen desorption (Fig. 5.11) [38].

Although having quite high kinetics of the above processes in the UFG state of the alloy, the application of ECAP and alloying with catalysts could not significantly reduce the hydrogen absorption/desorption temperature in ZK60 alloy. The temperatures remain too high for application in practice, for example, in automobile

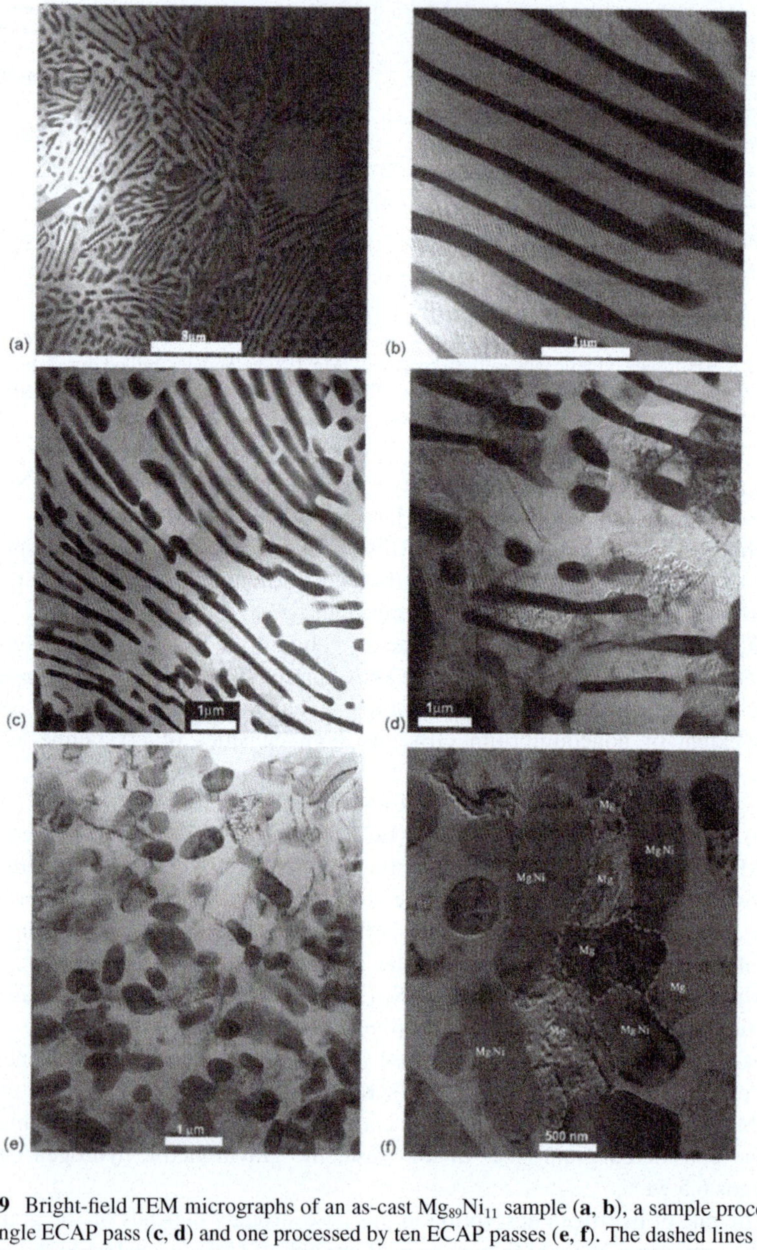

Fig. 5.9 Bright-field TEM micrographs of an as-cast $Mg_{89}Ni_{11}$ sample (**a**, **b**), a sample processed by a single ECAP pass (**c**, **d**) and one processed by ten ECAP passes (**e**, **f**). The dashed lines in (**f**) serve as a guide to the eye for recognizing the individual Mg grains and Mg_2Ni particles. (Reproduced from [37], Copyright 2009, with permission of John Wiley & Sons)

engines. In this regard, the search for hydrogen storage materials with lower hydrogen desorption temperatures remains relevant, although the capacity of such materials in relation to hydrogen may be lower than that of the ZK60 alloy. At the same

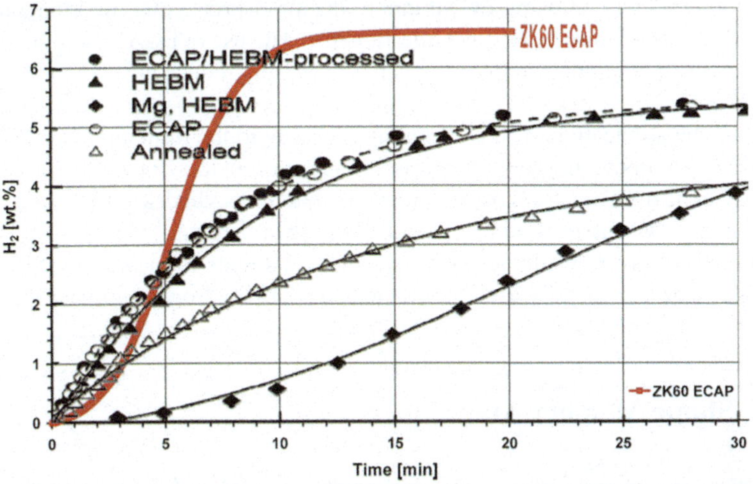

Fig. 5.10 Comparison of hydrogen desorption kinetics of nanostructured ZK60 alloy fabricated by HEBM, and/or by eight ECAP passes done at 300 °C with the final pass at 20 °C (black points and lines). The red line represents the kinetics of ZK60 after four ECAP passes done at steadily decreasing temperatures from 300 °C to 20 °C. The kinetics of both ball-milled, pure Mg, and the coarse-grained ZK60 alloy are also given. (Reproduced from [38], Copyright 2010, with permission of John Wiley & Sons)

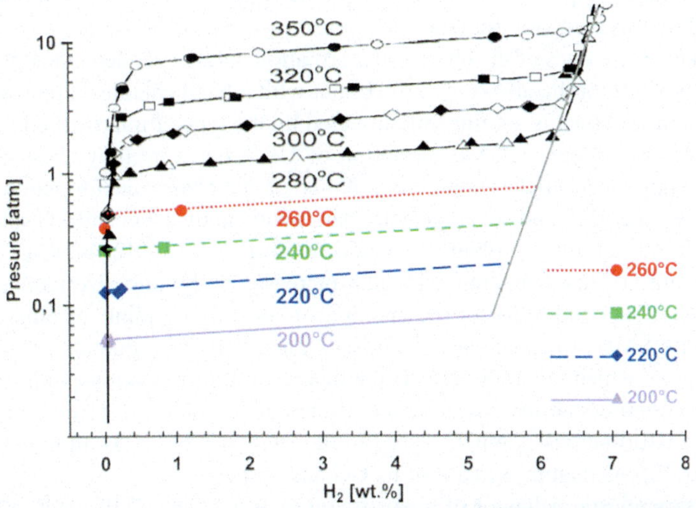

Fig. 5.11 Pressure-composition isotherms in ECAP- and/or HEBM-processed ZK60 alloy. The open and filled symbols indicate hydrogen absorption and desorption, respectively. The black points and lines are from ECAP/HEBM-processed ZK60, while the colored were measured on ECAP-ZK60. (Reproduced from [38], Copyright 2010, with permission of John Wiley & Sons)

time, the efficiency of using ECAP due to the short processing duration and relatively low cost of equipment is not questionable. The use of ECAP-Conform allows switching to a continuous process, which will sharply reduce the cost for obtaining UFG billets.

Thus, the application of SPD techniques leading to the formation of UFG states in Mg alloys opens up new prospects in creating devices for hydrogen storage. ECAP can be used as a low cost technique of obtaining cylindrical hydrogen battery cells, and ECAP can be used to create flat hydrogen batteries [39].

The above-given results suggest that UFG materials fabricated by SPD techniques can be successfully used for efficient and safe hydrogen storage [34, 40].

5.4 Shape Memory Alloys

The phenomenon and nature of the shape memory effect (SME) have been considered in Sect. 3.3. Superelasticity, which consists of the shape recovery of a strained workpiece after a martensitic phase transformation under load, was also discussed.

Shape memory effect and superelasticity are unique functional properties of great practical interest [41]. Major alloys exhibiting SME and superelasticity include NiTi nitinols, NiTi-based ternary alloys, and Cu-based alloys. A broad list of applications of the most well-known alloys with shape memory effect – nitinols, is presented, for example, on the NiTiMet Company website (www.niti-met.ru). There are also samples of semi-finished products and items made of nitinols manufactured by this company for engineering and medicine (Fig. 5.12).

The specific values of the phase transformation parameters depend on the chemical composition of the alloys with SME, but with a fixed chemical composition the shape memory effect is significantly affected by the type of microstructure.

As shown in Sect. 3.3, the formation of UFG states in alloys with SME and superelasticity leads to the suppression of martensitic phase transformation [42, 43]. As a result, the SME and superelasticity parameters in the UFG and CG states of the alloys differ, increasing with grain size decrease (Fig. 5.13) [42, 44, 45].

The value of reactive stresses, which increases with grain size reduction, provides favorable conditions for the creation of superstrong joints of pipes made of alloys with SME [46]. The small grain size hinders the reproduction and sliding of dislocations, which favorably affects the thermal stability of alloys with SME [47]. The temperature stability of numerous cycles of forward and reverse martensitic phase transformations is much higher in the case of UFG NiTi alloy obtained by ECAP at 425 °C than in the case of its CG analog [47].

The superelastic behavior of nanostructured alloys with SME obtained by SPD techniques is also significantly different from that of their well annealed CG counterparts. The strain curves presented in Fig. 5.14 demonstrate that a very high tensile strength (more than 2 GPa) and superelasticity are achieved in a NiTi alloy wire subjected to cold drawing treatment with a true strain rate of about 1 and

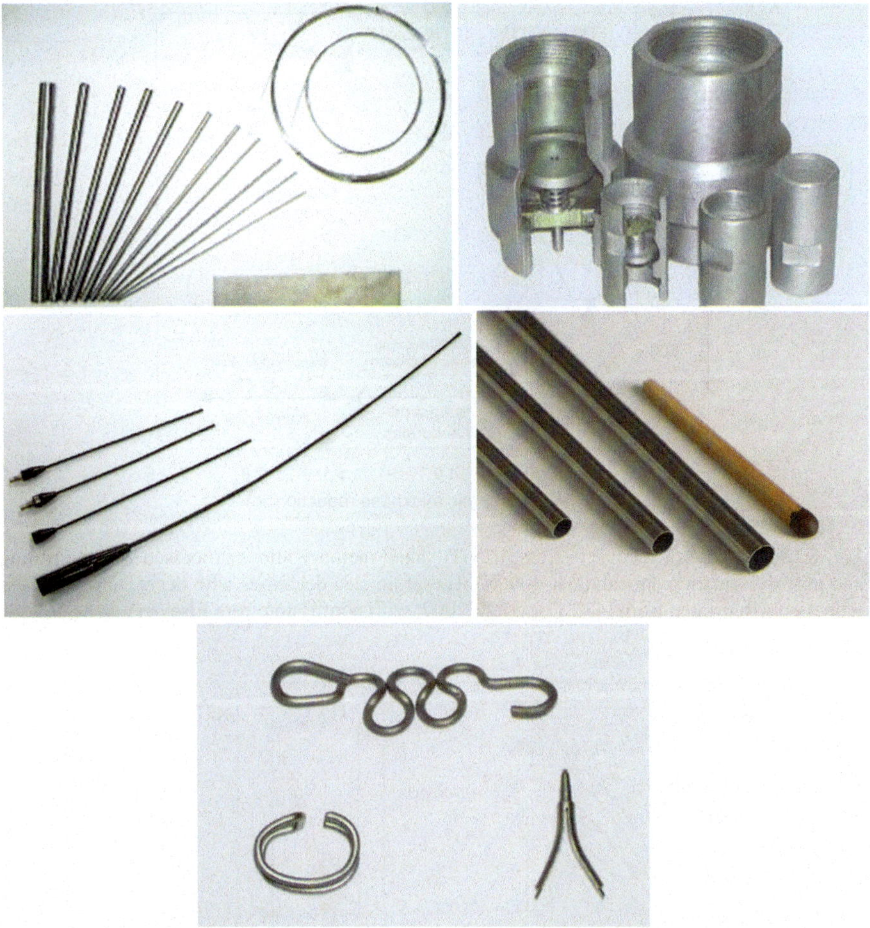

Fig. 5.12 Products made of Ni–Ti alloys with shape memory effect in the form of wire, sheets, pipes, rods, and ingots [41]. Source: http://www.niti-met.ru/niti.php. Author: Bondarev Andrey Borisovich

characterized by a microstructure consisting of a mixture of nanocrystalline and amorphous phases [48].

Under cyclic loading, the wires exhibit superelastic behavior with almost complete shape recovery exceeding 5%, very narrow hysteresis (measured as the area between the stress–strain curves obtained under load and unloading), and an apparent Young's modulus which increases significantly with enhancing degree of cold working (see inclusion in Fig. 5.14a). The change in the degree of relative compression during drawing affects the magnitude of superelasticity, which opens up a wide range of possibilities for innovative applications.

Thus, bulk UFG billets of the TiNi alloys with SME and superelasticity processed by SPD techniques demonstrate the combination of increased functional and mechanical properties [38, 49] and are promising for practical applications.

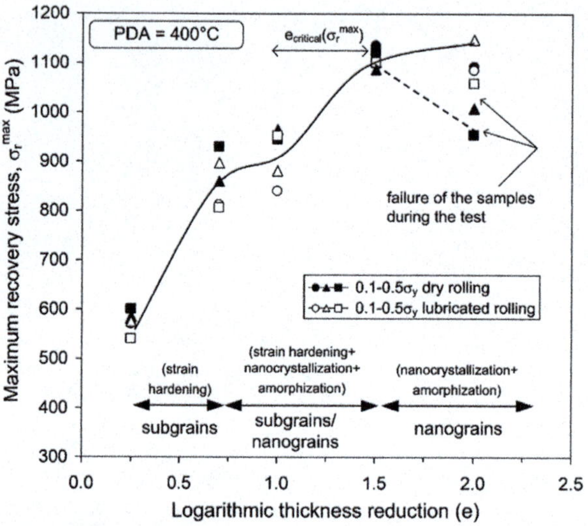

Fig. 5.13 Increase of recovery stress of NiTi shape memory alloys processed by cold rolling and post-deformation annealing at 400 °C. The grain size decreases with decreasing thickness reduction. (Reprinted from [45], Copyright 2009, with permission from Elsevier)

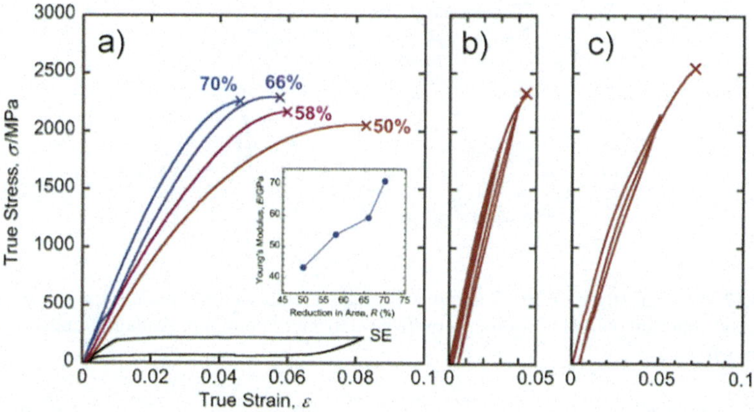

Fig. 5.14 Enhanced superelasticity of Ni-49.1 at.% Ti shape memory wires containing a mixture of nanocrystalline and amorphous phase [48]. Tensile-stress–strain curves obtained by cold drawing: (**a**) data for wires subjected to different degrees of drawing reduction (50, 58, 66, and 70%), and a commercial superplastic wire (SE). The Young's modulus versus drawing reduction is shown as an insert. The tensile strength and apparent elastic modulus of the drawn wires can exceed 2 and 70 GPa, respectively. (**b**) Cyclic-stress–strain curve of 70% drawn wire after aging for 3.6 ks at 573 K showing a strain recovery >5% (**c**). (Reprinted from [50], Copyright 2017, with permission from Elsevier)

Functionality and reliability are among the most important domains for tools in medicine. Manufacturing various medical devices using the UFG TiNi alloys offers a number of advantages, including a significant reduction in the size and mass, as

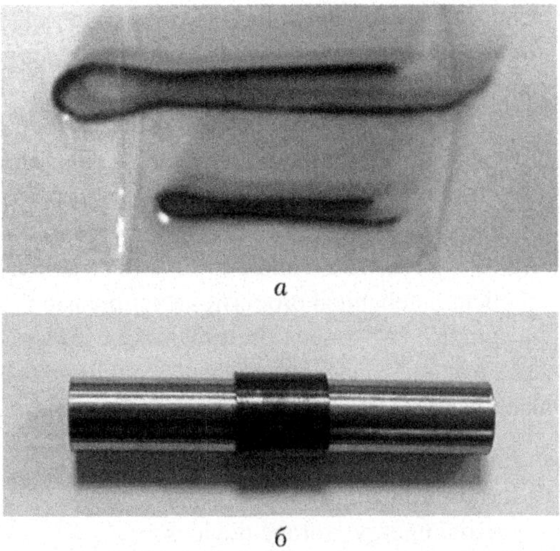

Fig. 5.15 A clip made of UFG TiNi for clipping blood vessels during laparoscopic surgeries (**a**) and a coupling for demountable thermomechanical connection of parts and structural elements (**b**). (Reprinted from [50], Copyright 2017, with permission from Elsevier)

well as an increase in their reliability due to the enhanced strength properties, power characteristics of SME, and superelasticity [38, 50]. A good example of practical application of UFG TiNi alloy is a device for clipping blood vessels, tubular structures, and soft elastic tissues that is designed to stop bleeding during laparoscopic surgeries. The UFG TiNi clip has twice the value of the maximum estimated force developed when the clip is actuated compared to a clip made of CG material (Fig. 5.15a).

Another example of the application of UFG alloys with SME is a coupling made of TiNi alloy with Fe additives, designed to provide increased tightness when joining pipelines and parts working under conditions of high pressure (Fig. 5.15b). Couplings made of Ti–Ni–Fe alloys are characterized by increased strength and tightness under the influence of axial loads, torque, and internal pressure.

The use of UFG materials with a unique set of service properties in medicine also contributes to the "social effect," which is realized in an increase in the quality of surgical interventions, reduction of their traumatism and, consequently, a significant reduction in the patient's rehabilitation period.

5.5 Nanostructured Magnets

In Sect. 3.2.2, it was demonstrated that SPD nanostructuring can dramatically improve the magnetic properties of soft- and hard-magnetic materials. Nanostructured $Re_2Fe_{14}B$ (Re = Nd, Pr)-based alloys with high coercivity H_C up to 2240 kA/m were produced via HPT followed by annealing at 600 °C [51]. Similar results were reported for melt-spun $Nd_9Fe_{85}B_6$ alloy [52]. Metal–ceramic powder mixtures consisting of Co (ferromagnetic) and NiO (antiferromagnetic) phases were compacted by HPT both with and without a prior ball milling (BM) [34].

Enhanced coercivity of the obtained materials was ascribed to the Co particle size refinement, the increase of stacking faults, and the magnetic coupling between Co and NiO phases [53].

Novel high-speed electrical machines require high-strength magnetic materials with ultimate fracture strength ≥900 MPa. However, the maximum ultimate fracture strength in conventional magnets processed via rolling or high-temperature forging does not exceed 700 MPa. Recently, it was demonstrated that nanostructured Fe–25Cr–15Co and Fe–30Cr–8Co alloys with good magnetic properties and excellent mechanical properties (compression stress of 1400 MPa) can be fabricated via complex two-stage upsetting torsion under isothermal conditions [54, 55]. These nanostructured magnetic materials could be used in developing high-speed electric machines for lower power consumption. These machines include the following:

- turbo-generators and turbo-compressors
- turbo-molecular pumps
- inertial energy-storage devices
- technological and medical centrifuges electric spindles and electric drivers

Another future application of nanostructured magnets is a generation of new material exhibiting energy waves, referred to as magnons, that coalesce into a common ground state in the manner of a Bose–Einstein condensate [56].

5.6 Micro- and Mini-devices from Ultrafine-Grained Metals

Micro- and mini-devices, with characteristic sizes from 10 mm to 10 micrometers, have the potential to be used in engineering, medicine, and everyday life. The image of a physical gear driven by a chain with the links being about 50 micrometers long, i.e., less than the diameter of a human hair serves as a striking example of a micro-device (Fig. 5.16) [57].

Fig. 5.16 The image of a physical gear driven by a chain with the links being about 50 micrometers long. (Reprinted from [57], Copyright 2022, with permission from Elsevier)

The creation of micro- and mini-devices comparable in size to miniature components of integrated electrical circuits opens up new opportunities for the design and development of microelectromechanical systems (MEMS) that combine microelectronic and micromechanical components and have their inherent advantages.

CG materials with a grain size of tens of micrometers cannot be used to create micro- and mini-devices due to the heterogeneity of the microstructure with a grain size comparable to the geometric size of micro- and mini-devices and the resulting anisotropy of the properties. This can be resolved by using UFG materials.

UFG materials have a grain size of tens to hundreds of nanometers, which is tens and hundreds of times less than the size of micro- and mini-devices. As a result, the problems of homogeneity and anisotropy of properties are resolved by using UFG materials. As considered earlier in this book, they obtain improved properties, such as strength, fatigue, wear resistance, magnetic properties, shape memory effect, superplasticity, corrosion resistance, and biocompatibility. At the same time, their level can be engineered by the choice of SPD regimes and subsequent deformation-heat treatment [39].

To date, the sizes of SPD-produced billets with UFG structure obtaining unique properties are usually small. However, they are quite sufficient to fabricate the parts for mini- and micro-devices. The laboratory equipment that allows the implementation of HPT or ECAP can be used for the production of UFG billets. In this case, each UFG billet, including an ECAP-subjected disk, a cylinder-shaped billet, or a bar billet with a square cross section, is sufficient to produce a batch of parts with improved properties for micro- and mini-devices [58].

A decrease in sample size can significantly affect the basic mechanisms of plasticity – in particular, the accumulation, generation, movement, and fixation of dislocations in conventional CG metallic materials. Simultaneously individual "dislocation avalanches" are often formed in microscopic samples resulting in an uncontrolled process of the development of localized deformation clusters, including the possibility of severe fracture. Thus, the manufacture of miniature components or devices, in particular components for biomedical devices or MEMS with geometric dimensions of several micrometers, requires a very precise scaling of their internal structures. In particular, the average grain size in microcomponents forming micro and mini-devices must be much smaller than the minimum size of the microcomponents.

The creation of UFG metallic materials opens up new opportunities in the fabrication of complex-shaped microcomponents for microdevices. For example, it has recently been shown that UFG stamping of Al 1050 alloy, obtained by ECAP, demonstrates high potential for use in the fabrication of microheat exchangers with geometric dimensions less than 10 μm and high thermal conductivity [59]. The stamping of this material provides a very high-quality defect-free surface of the finished product with a precise geometric shape, while the use of conventional hot stamping of Al 1050 alloy leads to a product with a coarse finish surface due to the formation of shear bands [59].

Unconventional application of the principles of ECAP processing has also been considered recently for the process of solid-state infiltration [60]. In this process, the penetration of aluminum into porous iron billets under applied high pressure is

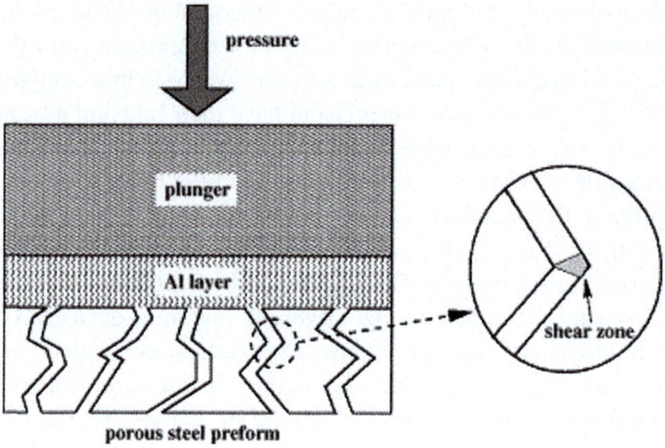

Fig. 5.17 The principle of force filling, similar to ECAP, as applied to solid infiltration. An enlarged image of an area bounded by a dotted circle, corresponding to the shear zones formed near the bends is given in the right image. (Reprinted from [60, 62], Copyright 2005, with permission from Yuri Estrin)

similar to the force-filling process used in microelectronics [61]. During plastic flow through sinuous "channels" inside a porous iron billet, aluminum repeatedly crosses areas where strong shear deformation occurs in much the same way as in ECAP (Fig. 5.17) [62]. The penetration of aluminum into the channels of the porous billet during solid-state infiltration is confirmed by the results presented in Fig. 5.18 [62].

HPT can be combined with the extrusion process. As a result, the required axisymmetrical profile of the final product can be obtained. A long billet with a saw-like profile made of AZ31 magnesium alloy serves as an example of a possible product [63]. Gears for MEMS can easily be cut from the resulting billet (Fig. 5.19) [63].

The above-presented and other examples [37, 64] prove that the use of UFG materials obtained by SPD techniques in the creation of micro and mini-devices for various purposes to be quite promising.

Thus, the results of this chapter provide a clear demonstration that processing by SPD can produce superior properties in bulk nanostructured materials. Based on the evidence available to date, significant information now exists on the underlying mechanisms that account for these special properties, opening new possibilities for practical applications of these materials in novel structural and functional applications. Recent discoveries demonstrate new opportunities for SPD processing for improving other physical and chemical properties of these materials. These properties were mentioned only briefly in this review but nevertheless have achieved considerable advances in recent years. These include, for example, increased superconductivity, thermoelectricity, giant magnetoresistance, improved hydrogen storage, and improved biocompatibility. Lastly, it's noteworthy that various SPD techniques are utilized in processing battery materials, resulting in enhanced properties for batteries based on Zn, Li, Mg, and Al [65]. This underscores another potential application of bulk nanostructured metals in our future.

A general rule in modern materials science is that any material breakthrough in the twentieth century required, from the time of the initial innovation, about two

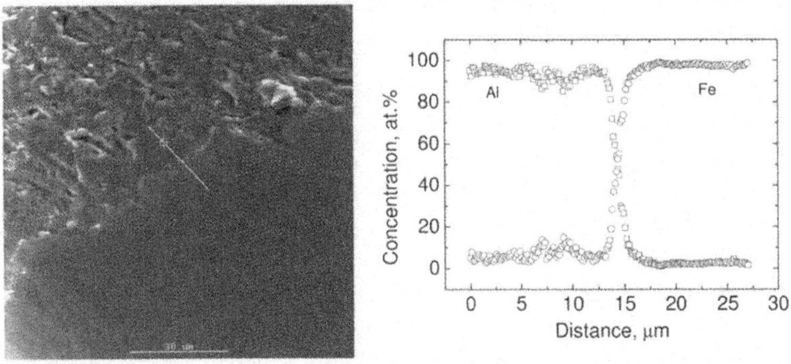

Fig. 5.18 Al and Fe concentration profiles across the interface measured by EDS along the line in the left picture. (Reprinted from [62], Copyright 2005, with permission from Yuri Estrin)

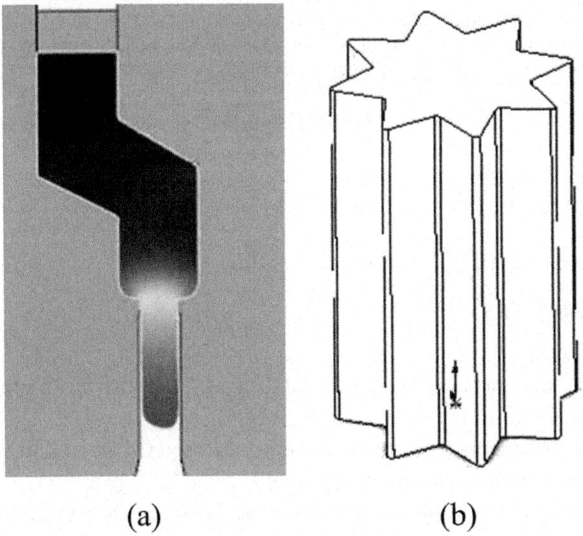

Fig. 5.19 Integrated ECAP/extrusion process: (**a**) die design combining two ECAP passes (channel angle 120°) and a final extrusion step; (**b**) possible product profile produced in the exit channel of the die. (Reprinted by permission from Springer Nature [57], Copyright 2007)

decades to achieve widespread market acceptance. It appears that bulk nanostructured metallic materials, as produced by SPD processing, are also following this track. Although the first developments and research started at the beginning of the 1990s, there have been very significant developments in the commercialization of these materials in recent years. This is especially evident from the widespread production of advanced pilot articles with new functionalities.

There are three primary advantages associated with the application and commercialization of bulk nanostructured metals: markedly superior properties, the potential for efficient fabrication using SPD-processing techniques such as ECAP-C, and the possibility of using these materials to produce cutting-edge products. Furthermore, many of these new applications involve, or will involve, extreme environmental conditions requiring both exceptional strength and improved functional properties.

It is reasonable to anticipate that, in the very near future, the nanostructuring of materials by SPD processing will lead to new breakthroughs in the development of bulk solids with superior properties for advanced structural and functional applications.

Suggested Reading

1. Valiev RZ, Sabirov I, Zhilyaev AP, Langdon TG (2012) Bulk nanostructured metals for innovative applications. JOM 64:1134–1142
2. Lowe TC, Valiev RZ, Li X, Ewing B (2021) Commercialization of bulk nanostructured metals and alloys. MRS Bull 46:265–272
3. Valiev RZ, Estrin Y, Horita Z, Langdon TG, Zehetbauer MJ, Zhu YT (2016) Producing bulk ultrafine-grained materials by severe plastic deformation: ten years later. JOM 68:1216–1226
4. Estrin Y, Zehetbauer MJ (2009) Niche applications of bulk nanostuctured materials processed by severe plastic deformation. In: Zehetbauer MJ, Zhu YT (eds) Bulk nanostructured materials. WILEY-VCH Verlag GmbH &Co. KGaA, Wienheim, pp 635–648
5. Valiev RZ, Zhilyaev AP, Langdon TG (2014) Bulk nanostructured materials: fundamentals and applications. Wiley/TMS, Hoboken
6. Valiev RZ, Estrin Y, Horita Z, Langdon TG, Zehetbauer MJ, Zhu YT (2016) Fundamentals of superior properties in bulk NanoSPD materials. Mater Res Lett 4:1–21

Problems and Tasks

1. Reason the advantages of using implants produced from UFG titanium and titanium alloys in medicine.
2. Analyze the prospects for the use of nanostructured Cu and Al alloys as materials for electrical industry and electronics.
3. Make a brief presentation as a promoter of UFG metal materials for hydrogen storage.
4. What are the prospects for the use of UFG alloys with shape memory effect?
5. What is the relationship between the microstructural features, specific enhanced properties of the prospects for the use of nanostructured magnetic materials?
6. What microstructure parameters open prospects for the use of UFG materials for manufacturing of micro and mini-devices?

References

1. Valiev RZ, Sabirov I, Zhilyaev AP, Langdon TG (2012) Bulk nanostructured metals for innovative applications. JOM 64:1134–1142
2. Semenova IP, Valiev RZ, Smyslov AM, Pesin MV, Langdon TG (2021) Advanced materials for mechanical engineering: ultrafine-grained alloys with multilayer coatings. Adv Eng Mater 23:2100145

3. Lowe TC, Valiev RZ, Li X, Ewing B (2021) Commercialization of bulk nanostructured metals and alloys. MRS Bull 46:265–272
4. Lowe TC (2006) Metals and alloys nanostructured by severe plastic deformation: commercialization pathways. JOM 58:28
5. Hanawa T (2012) Research and development of metals for medical devices based on clinical needs. Sci Technol Adv Mater 13:064102
6. Hanawa T (2010) Metals for biomedical devices. Woodhead Publishing Limited, Oxford, p 324
7. Froes FH, Qian M (2018) Titanium in medical and dental applications. Woodhead Publishing, Duxford, p 630
8. Valiev RZ, Islamgaliev RK, Alexandrov IV (2000) Bulk nanostructured materials from severe plastic deformation. Prog Mater Sci 45:103189
9. Valiev RZ, Estrin Y, Horita Z, Langdon TG, Zehetbauer MJ, Zhu YT (2016) Producing bulk ultrafine-grained materials by severe plastic deformation: ten years later. JOM 68:1216–1226
10. Valiev RZ, Sabirov I, Zemtsova EG, Parfenov EV, Dluhoš L, Lowe TC (2018) In: Froes F, Qian M (eds) Titanium in medical and dental applications. Woodhead Publishing, Duxford, pp 393–418
11. Semenova IP, Klevtsov GV, Klevtsova NA, Dyakonov GS, Matchin AA, Valiev RZ (2016) Nanostructured titanium for maxillofacial mini-implants. Adv Eng Mater 18:12161224
12. ISO 14602 Non-active surgical implants-Implants for Osteosynthesis particular requirements
13. http://www.instron.com/en/testing-solutions/by-test-type/simple-cyclic/orthopaedic-micro-implants
14. Brunette DM, Tengvall P, Textor M, Thomsen P (2003) Titanium in medicine. Springer, Berlin/Heidelberg
15. Polyakov AV, Dluhos L, Dyakonov GS, Raab GI, Valiev RZ (2015) Recent advances in processing and application of nanostructured titanium for dental implants. Adv Eng Mater 17:1869–1875
16. Nie FL, Zheng YF, Wei SC, Wang DS, Yu ZT, Salimgareeva GK, Polyakov AV, Valiev RZ (2013) In vitro and in vivo studies on nanocrystalline Ti fabricated by equal channel angular pressing with microcrystalline CP Ti as control. J Biomed Mater Res Part A 101:1694–1707
17. Bagherifard S, Ghelichi R, Khademhosseini A, Guagliano M (2014) Cell response to nanocrystallized metallic substrates obtained through severe plastic deformation. ACS Appl Mater Interfaces 6:7963–7985
18. Polyakov AV, Dyakonov GS, Semenova IP, Raab GI, Dluhos L, Valiev RZ (2015) Development and study of medical implants made from nanostructured titanium. Adv Biomater Dev Med 2:63–69
19. Zemtsova EG, Arbenin AY, Valiev RZ, Orekhov EV, Semenov VG, Smirnov VM (2016) Two-level micro-to-nanoscale hierarchical TiO_2 nanolayers on titanium surface. Materials 9:1010
20. Shin MH, Baek SM, Polyakov AV, Semenova IP, Valiev RZ, Hwang WB, Hahn SK, Kim HS (2018) Sci Rep 8:9907
21. Parfenov EV, Parfenova LV, Dyakonov GS, Danilko KV, Mukaeva VR, Farrakhov RG, Lukina ES, Valiev RZ (2019) Surf Coat Technol 357:669683
22. Nazarov DV, Zemtsova EG, Solokhin AY, Valiev RZ, Smirnov VM (2017) Modification of the surface topography and composition of ultrafine and coarse grained titanium by chemical etching. Nano 7:15
23. Valiev RZ, Parfenov EV, Parfenova LV (2019) Developing nanostructured metals for manufacturing of medical implants with improved design and biofunctionality. Mater Trans 60:1356–1366
24. Rajagopalan M, Darling KA, Kale C, Turnage SA, Koju RK, Hornbuckle BC, Mishin Y, Solanki KN (2019) Nanotechnology enabled design of a structural material with extreme strength as well as thermal and electrical properties. Mater Today 31:10–20
25. Kuhn HA, Altenberger I, Käufler A, Hölzl H, Fünfer M (2012) Properties of high performance alloys for electromechanical connectors. In: Collini L (ed) Copper alloys – early applications and current performance – enhancing processes. InTech Publisher, Rijeka, pp 51–68

26. Wei KX, Chu ZQ, Yang LC, Wei W, Du QB, Alexandrov IV, Hu J (2020) Performance evaluation of electrical discharge machining using ultra-fine grained Cu electrodes by ECAP and DCT. J Mater Eng Perform 30:281–289
27. Sokoluk M, Cao C, Pan S, Li X (2019) Nanoparticle-enabled phase control for arc welding of unweldable aluminum alloy 7075. Nat Commun 10:1
28. Ding C, Xu J, Shan D, Guo B, Langdon TG (2021) Sustainable fabrication of Cu/Nb composites with continuous laminated structure to achieve ultrahigh strength and excellent electrical conductivity. Compos Part B 211:108662
29. Murashkin MY, Sabirov I, Sauvage X, Valiev RZ (2016) Nanostructured Al and Cu alloys with superior strength and electrical conductivity. J Mater Sci 51:33–49
30. Zhang X, Keramati H, Arie M, Singer F, Tiwari R, Shooshtari A, Ohadi M (2018) Recent developments in high temperature heat exchangers: a review. Front Heat Mas Trans 11:18
31. Mann VKh, Krokhin AYu, Matveeva IA, Raab GI, Murashkin MYu, Valiev RZ (2014) Nanostructured wire rod research and development in Russia. Light Metal Age 72:32–34
32. Raab GJ, Valiev RZ, Lowe TC, Zhu YT (2004) Continuous processing of ultrafine grained Al by ECAP – conform. Mater Sci Eng A 382:30–34
33. Varin RA, Czujko T, Wronski ZS (2009) Nanomaterials for solid state hydrogen storage. Springer, New York
34. Leiva DR, Jorge AM Jr, Ishikawa TT, Botta WJ (2019) Hydrogen storage in Mg and Mg-based alloys and composites processed by severe plastic deformation. Mater Trans 60:1561–1570
35. Zaluski L, Zaluska A, Tessier P, Ström-Olsen JO, Schulz R (1996) Nanocrystalline hydrogen absorbing alloys. Mater Sci Forum 225:853
36. Révész Á, Kánya Z, Verebélyi T, Szabó PJ, Zhilyaev AP, Spassov T (2010) The effect of high-pressure torsion on the microstructure and hydrogen absorption kinetics of ball-milled $Mg_{70}Ni_{30}$. J Alloys Compd 504:83
37. Estrin Y, Zehetbauer MJ (2009) Niche applications of bulk nanostuctured materials processed by severe plastic deformation. In: Zehetbauer MJ, Zhu YT (eds) Bulk nanostructured materials. Wiley-VCH Verlag GmbH &Co. KGaA, Wienheim, pp 635–648
38. Zehetbauer M, Grössinger R, Krenn H, Krystian M, Pippan R, Rogl P, Waitz T, Würschum R (2010) Bulk nanostructured functional materials by severe plastic deformation. Adv Eng Mater 12:692–700
39. Valiev RZ, Zhilyaev AP, Langdon TG (2014) Bulk nanostructured materials: fundamentals and applications. Wiley/TMS, Hoboken
40. Valiev RZ, Estrin Y, Horita Z, Langdon TG, Zehetbauer MJ, Zhu YT (2016) Fundamentals of superior properties in bulk NanoSPD materials. Math Res Lett 4:1–21
41. Otsuka K, Wayman CM (1999) Shape memory materials. Cambridge University Press, Cambridge, MA, p 284
42. Waitz T, Tsuchiya K, Antretter T, Fischer FD (2009) Phase transformations of nanocrystalline martensitic materials. MRS Bull 34:11
43. Sergueeva AV, Song C, Valiev RZ, Mukherjee AK (2003) Structure and properties of amorphous and nanocrystalline NiTi prepared by severe plastic deformation and annealing. Mater Sci Eng A 339:159
44. Waitz T, Kazykhanov V, Karnthaler HP (2004) Martensitic phase transformations in nanocrystalline NiTi studied by TEM. Acta Mater 52:137
45. Demers V, Brailovski V, Prokoshkin SD, Inaekyan KE (2009) Optimization of the cold rolling processing for countinous manufacturing of nanostructured TiNi shape memory alloys. J Mater Process Technol 209:3096
46. Waitz T, Antretter T, Fischer FD, Karnthaler HP (2008) Size effects on martensitic phase transformations in nanocrystalline NiTi shape memory alloys. Mater Sci Technol 24:934
47. Churakova AA, Gunderov DV (2020) Structural features of nanostructural equiatom alloy TiNi during multiple martensite transformations. Nano 13:44–53
48. Tsuchiya K, Hada Y, Koyano T, Nakajima K, Ohnuma M, Koike T, Todaka Y, Umemoto M (2009) Production of TiNi amorphous/nanocrystalline wires with high strength and elastic modulus by severe cold drawing. Scr Mater 60:749

49. Valiev R, Gunderov D, Prokofiev E, Pushin V, Zhu Y (2008) Nanostruturing of TiNi alloy by SPD processing for advanced properties. Mater Trans 49:97–101
50. Ryklina E, Korotitskiy A, Khmelevskaya I, Prokoshkin S, Polyakova K, Kolobova A, Soutorine M, Chernov A (2017) Control of phase transformations and microstructure for optimum realization of one-way and two-way shape memory effects in removable surgical clips. Mater Des 136:174–184
51. Stolyarov VV, Gunderov DV, Valiev RZ, Popov AG, Gaviko VS, Ermolenko AS (1999) Metastable nanostructured states in $Re_2Fe_{14}B$-based alloys processed by severe plastic deformation. J Magn Magn Mater 196:166–168
52. Popov AG, Gaviko VS, Shchegoleva NN, Shreder LA, Gunderov DV, Stolyarov VV, Li W, Li LL, Zhang XY (2006) Effect of high-pressure torsion deformation and subsequent annealing on structure and magnetic properties of overquenched melt-spun $Nd_9Fe_{85}B_6$ alloy. J Iron Steel Res Int 13:160–165
53. Menéndez E, Sort J, Langlais V, Zhilyaev A, Muñoz JS, Suriñach S, Nogués J, Baró MD (2007) Cold compaction of metal–ceramic (ferromagnetic–antiferromagnetic) composites using high pressure torsion. J Alloys Compd 434–435:505–508
54. Korznikova GF, Korznikov AV (2009) Gradient submicrocrystalline structure in Fe–Cr–Co system hard magnetic alloys. Mater Sci Eng A 503:99–102
55. Korneva A, Bieda M, Korznikova G, Sztwiertnia K, Korznikov A (2008) Microstructure and some properties of $FeCr_{25}Co_{15}$ alloy subjected to plastic deformation by complex load. J Mater Res 99:991–998
56. Kittel C (2005) Introduction to solid state physics, 8th edn. Wiley, p 704
57. Marsh G (2002) Micromachinery rolling at last? Mater Today 5:44–49
58. Estrin Y, Kulagin R, Beygelzimer Y (2018) Think big, manufacture small: microfabrication on a desktop by severe plastic deformation. Mater Today
59. Qiao XG, Gao N, Moktadir Z, Kraft M, Starink MJ (2010) Fabrication of MEMS components using ultrafine-grained aluminium alloys. J Micromech Microeng 20:045029
60. Estrin Y, Rabkin E, Hellmig RJ, Kazakevich M, Zi A (2005) Severe plastic deformation by solid state infiltration. Mater Sci Eng A 410–411:165
61. Estrin Y, Kim HS, Kovler M, Berler G, Shaviv R, Rabkin E (2003) Modeling of aluminum via filling by forcefill. J Appl Phys 93:5812
62. Estrin Y, Rabkin E, Hellmig RJ, Kazakevich M, Zi A (2005) Solid state infiltration of porous steel with aluminium by the forcefill process. Int J Mater Res 96:1193
63. Estrin Y, Janecek M, Raab GI, Valiev RZ, Zi A (2007) Severe plastic deformation as a means of producing ultra-fine-grained net-shaped micro electro-mechanical systems parts. Metal Mater Trans A38:1906
64. Valiev RZ, Zehetbauer MJ, Estrin Y, Höppel HW, Ivanisenko Y, Hahn H, Wilde G, Roven HJ, Sauvage X, Langdon TG (2007) The innovation potential of bulk nanostructured materials. Adv Eng Mater 9:527
65. Zohrevand M, Hassanzadeh N, Alizadeh R et al. (2023) Review: recent advances using severe plastic deformation to improve the properties of battery materials. J Mater Sci (2023). https://doi.org/10.1007/s10853-023-09198-5

Concluding Remarks

Scientific and technological progress is impossible without the development and extensive use of new promising materials with enhanced properties and industrial technologies that allow obtaining materials and semi-finished products (threads, wires, rods, bands, sheets, etc.) as well as products made from them. The material presented in this textbook convincingly demonstrates the important niche occupied by SPD-produced UFG metallic materials and composites among new promising materials. The knowledge about the techniques of production of UFG materials, their microstructural features, enhanced mechanical properties, favorable functional properties, and prospects of application in industry will help students of various fields of fundamental (physicists, mechanics, chemists), engineering (mechanical engineers, power engineers, transport workers, etc.), medical (traumatologists, dentists) programs to face new challenges in manufacturing, electronics, energy, and medicine and provide a major qualitative and quantitative step-change in science for a more sustainable future.

R. Z. Valiev et al., *Ultrafine-Grained Materials*, The Minerals, Metals & Materials Series, https://doi.org/10.1007/978-3-031-31729-3

Index

R. Z. Valiev et al., *Ultrafine-Grained Materials*, The Minerals, Metals & Materials Series, https://doi.org/10.1007/978-3-031-31729-3

The manufacturer's authorised representative in the EU is Springer Nature Customer Service Centre GmbH, Europaplatz 3, 69115 Heidelberg, Germany. If you have any concerns regarding our products, please contact ProductSafety@springernature.com

Printed and bound by CPI Group (UK) Ltd, Croydon, CR0 4YY

01/10/2026

02225883-0013